Time Series Analysis
in Meteorology
and Climatology

Companion Website

This book has a companion website www.wiley.com/go/duchon/timeseriesanalysis with datasets for the problems in each chapter.

Time Series Analysis in Meteorology and Climatology

An Introduction

Claude Duchon

University of Oklahoma, USA

Robert Hale

Colorado State University, USA

WILEY-BLACKWELL

A John Wiley & Sons, Ltd., Publication

This edition first published 2012 © 2012 by John Wiley & Sons, Ltd.

Wiley-Blackwell is an imprint of John Wiley & Sons, formed by the merger of Wiley's global Scientific, Technical and Medical business with Blackwell Publishing.

Registered office: John Wiley & Sons, Ltd, The Atrium, Southern Gate, Chichester, West Sussex, PO19 8SQ, UK

Editorial offices: 9600 Garsington Road, Oxford, OX4 2DQ, UK
           The Atrium, Southern Gate, Chichester, West Sussex, PO19 8SQ, UK
           111 River Street, Hoboken, NJ 07030-5774, USA

For details of our global editorial offices, for customer services and for information about how to apply for permission to reuse the copyright material in this book please see our website at www.wiley.com/wiley-blackwell.

Library of Congress Cataloging-in-Publication Data

Duchon, Claude E.
   Time series analysis in meteorology and climatology : an introduction / Claude Duchon and Robert Hale.
      p. cm. – (Advancing weather and climate science)
   Includes bibliographical references and index.
   ISBN 978-0-470-97199-4 (hardback)
  1. Meteorology–Statistical methods. 2. Time-series analysis. I. Hale, Robert, 1940- II. Title.
   QC874.5.D83 2011
   551.501'51955–dc23

                                                  2011030428

A catalogue record for this book is available from the British Library.

This book is published in the following electronic formats: ePDF 9781119953098; Wiley Online Library 9781119953104; ePub 9781119960980; Mobi 9781119960997

Set in 10.5/13pt Minion by Thomson Digital, Noida, India
Printed and bound in Singapore by Markono Print Media Pte Ltd

First Impression 2012

Contents

Series foreword

Advances in weather and climate

Meteorology is a rapidly moving science. New developments in weather forecasting, climate science and observing techniques are happening all the time, as shown by the wealth of papers published in the various meteorological journals. Often these developments take many years to make it into academic textbooks, by which time the science itself has moved on. At the same time, the underpinning principles of atmospheric science are well understood but could be brought up to date in the light of the ever increasing volume of new and exciting observations and the underlying patterns of climate change that may affect so many aspects of weather and the climate system.

In this series, the Royal Meteorological Society, in conjunction with Wiley–Blackwell, is aiming to bring together both the underpinning principles and new developments in the science into a unified set of books suitable for undergraduate and postgraduate study as well as being a useful resource for the professional meteorologist or Earth system scientist. New developments in weather and climate sciences will be described together with a comprehensive survey of the underpinning principles, thoroughly updated for the 21st century. The series will build into a comprehensive teaching resource for the growing number of courses in weather and climate science at the undergraduate and postgraduate levels.

Series Editors

Peter Inness
University of Reading, UK

William Beasley
University of Oklahoma, USA

Preface

Time series analysis is widely used in meteorological and climatological studies because the vast majority of observations of atmospheric and land surface variables are ordered in time (or space). Over the years we have found a continuing interest by both students and researchers in our profession (and those allied to it) in understanding basic methods for analyzing observations ordered in time or space and evaluating the results. The purpose of this book is to respond to this interest. We've done this by deriving and interpreting various equations that are useful in explaining the structure of data and then, using computer programs, applying them to meteorological data sets. Overall, the material we cover serves as an introduction in the application of statistics to the analysis of univariate time series. The topics discussed should be relevant to anyone in any science where events are observed in time and/or space. To demonstrate a procedure, we use scalar atmospheric variables, for example, air temperature. Anyone who completes the five chapters, including working the problems at the end of each chapter, will have acquired sufficient understanding of time series terminology and methodology to confidently deal with more advanced spectrum analysis, for example, that found in radar and atmospheric turbulence measurements, analysis, and theory.

Chapter 1 deals with Fourier analysis and is divided into five sections. In the first three sections, mathematical formulas for representing a time series by Fourier sine and cosine coefficients are developed and their inherent symmetry emphasized. These formulas are applied to three data sets, two of which are actual observations. The three sections provide the background necessary to apply Fourier analysis to a time series, and one of the end-of-chapter problems invites the reader to write a computer program designed to accomplish this.

In the fourth section of Chapter 1 we investigate statistical properties of the Fourier spectrum. These statistical properties arise because time series from the physical world are usually nondeterministic, that is, no two data sets are alike. We explore the concept of a random variable, a realization, a population, stationarity, expectation, and a probability density function. The goal is to understand how random data produce a distribution of variances at each harmonic frequency and the statistical properties of this distribution. Armed with this information, the last part of this section involves testing the hypothesis that a particular data set, as viewed through the Fourier spectrum, is a sample from a population of white noise, that is, random numbers.

The fifth section of Chapter 1 is an examination of various topics relevant to time series analysis. We discuss aliasing, spectrum folding, and spectrum windows, phenomena that are a direct consequence of digital sampling, and show examples of each. In addition, we develop the Fourier transform, the mathematical formula that in one step converts a time series into its Fourier components in the frequency domain.

Chapter 1 is the longest of the five chapters because it encompasses both theory and application of Fourier analysis, relevant statistical concepts, and the foundation of methods of time series analysis developed in the remaining chapters.

The subject of Chapter 2 is linear systems. This chapter is the study of the relationships between two time series, an input series and an output series, and the associated input and output spectra. What links the two time series is a physical system, as in the case of measurement of some physical variable (for example, a thermometer to measure temperature), or a mathematical system, as in the case of filtering an observed time series to remove unwanted noise in the data.

Fundamental to Chapter 2 is the convolution integral. Whether a system is physical or mathematical, the convolution integral provides the mathematical connection between the input and output series, and its Fourier transform provides the connection between the input and output spectra.

Most variables of interest in the physical sciences are continuous in time (or space). Nevertheless, we practically always analyze digital time series. We investigate the relationship between analog and digital time series using a generalized function called the Dirac delta function. Through its application we can explain how the structure of an output time series that has passed through a linear system is altered relative to the input time series in terms of modified Fourier coefficients and phase angles. Two examples are discussed, a first order linear system and an integrator, both of which have practical use in meteorology and climatology, and the physical sciences in general.

Chapter 3 is principally about nonrecursive data filtering; that is, a filtered output time series is related only to the input time series – there is no feedback (as in recursive filtering). Time series that are to be filtered are viewed as data that already have been collected as opposed to real time filtering.

The primary objective of this chapter is to design and apply a two-parameter filter called the Lanczos filter. The two design parameters are the number of weights and the frequency that separates the Fourier spectrum into harmonic variances that remain unchanged and those that are suppressed. This filter provides its designer much more control of the filtering process than simple one-parameter filters, for example, the running mean. The theory of Lanczos filtering is developed, examples of its use are shown, and a computer program is provided so that the reader can apply the procedure to a data set.

One of the goals of a physical scientist is to understand the morphology of natural events. An obvious step that must be taken is to obtain samples in time and/or space of variables that characterize the physical properties of an event over its lifetime. The fact that an event has a lifetime implies that it evolves in time and/or space, a consequence of which is that successive observations of its properties are related. This is called autocorrelation, the title of Chapter 4. To realize the importance of autocorrelation in analyzing time series, we compare the formula for calculating the variance of the mean of a random variable with autocorrelation to that without autocorrelation. The latter formula is the form seen in typical undergraduate statistics texts while the former formula takes into account the degree of dependence in the time series.

In Chapter 4 we are interested in finding the best formula for estimating the mean, variance, autocovariance function, and autocorrelation function of a population of time series based, typically, on a single observed time series taken from that population. We examine populations of independent as well as autocorrelated data. Among the five chapters, this one is the most statistically oriented.

The lagged-product method discussed in Chapter 5 is an alternative to Fourier analysis. Quite often, Fourier analysis of geophysical data yields noisy-looking spectra. When this occurs, it is common to smooth a spectrum to make it more visually interpretable. In the lagged-product method, a smoothed variance spectrum can be obtained directly from the Fourier transform of the product of the auto-covariance function with another function that alters its shape. The degree of smoothing is controlled entirely by the latter function. The term lagged-product is used because the autocovariance function comprises time-lagged (or spatially-lagged) products and it is the autocovariance function that is being transformed.

This book was written for students and scientists who have a background in calculus and statistics, and familiarity with complex variables. Prior in-depth study of complex variables is not required.

The authors wish to thank the many students who have provided valuable comments and corrections over the years the material was used as lecture notes. Chapters 2, 4, and 5 were inspired by the book *Spectral Analysis and its Applications* (1968) by G.M. Jenkins and D.G. Watts, a classic volume in time series analysis.

Claude Duchon and Rob Hale
22 May 2011

1

Fourier analysis

It is often the case in the physical sciences, and sometimes the social sciences as well, that measurements of a particular variable are collected over a period of time. The collected values form a data set, or *time series*, that may be quite lengthy or otherwise difficult to interpret in its raw form. We then may turn to various types of statistical analyses to aid our identification of important attributes of the time series and their underlying physical origins. Basic statistics such as the mean, median, or total variance of the data set help us succinctly portray the characteristics of the data set as a whole, and, potentially, compare it to other similar data sets.

Further insight regarding the time serics, however, can be gained through the use of *Fourier, harmonic,* or *periodogram analysis* – three names used to describe a single methodology. The primary aim of such an analysis is to determine how the total variance of the time series is distributed as a function of frequency, expressed either as ordinary frequency in cycles per unit of time, for example, cycles per second, or angular frequency in radians per unit of time. This allows us to quantify, in a way that the basic statistics named above cannot, any *periodic* components present in the data. For example, outside air temperature typically rises and falls with some regularity over the course of a day, a periodic component governed by the rising and setting of the sun as the earth rotates about its axis. Such a periodic component is readily apparent and quantifiable after applying Fourier analysis, but is not described well by the mean, median, or total variance of the data.

In the first two sections of Chapter 1, we will learn some essential terminology of Fourier analysis and the fundamentals of performing Fourier analysis and its inverse, Fourier synthesis. Example data sets and their analyses are presented in Section 1.3 to further aid in understanding the methodology.

As with other types of statistical analyses, statistical significance plays an important role in Fourier analysis. That is, after performing a Fourier analysis, what if we

Time Series Analysis in Meteorology and Climatology: An Introduction, First Edition. Claude Duchon and Robert Hale.
© 2012 John Wiley & Sons, Ltd. Published 2012 by John Wiley & Sons, Ltd.

find that the variance at one frequency is noticeably larger than at other frequencies? Is this the result of an underlying physical phenomenon that has a periodic nature? Or, is the larger variance simply statistical chance, owing to the random nature of the process? To answer these questions, in Section 1.4 we examine how to ascribe confidence intervals to the results of our Fourier analysis.

In Section 1.5, we take a more detailed look at particular issues that may be encountered when using Fourier analyses. Although not generally requisite to performing a Fourier analysis, the concepts covered are often critical to correct interpretation of the results, and in some cases may increase the efficacy of an analysis. An understanding of these topics will allow an investigator to pursue Fourier analysis with a high degree of confidence.

1.1 Overview and terminology

1.1.1 Obtaining the Fourier amplitude coefficients

The goal of Fourier analysis is to decompose a data sequence into harmonics (sinusoidal waveforms) such that, when added together, they reproduce the time series. What makes sinusoidal waveforms an appropriate representation of the data is their orthogonality property, their ability to successfully model waves in the atmosphere, oceans, and earth, as well as phenomena resulting from solar forcing, and the fact that the harmonic amplitudes are independent of time origin and time scale (Bloomfield, 1976, p. 7).

Harmonic frequencies are gauged with respect to the *fundamental period*, the shortest record length for which the time series is not repeated. In most practical cases, this is the entire length of the available record, since the record typically does not contain repeated sequences of identical data. The harmonic frequencies include harmonic 1, which corresponds to one cycle over the fundamental period, and higher harmonics that are integer multiples of one cycle. Thus each harmonic is always an integer number of cycles over the length of the funda- mental period.

To establish a sense of Fourier analysis, consider a simple example. The heavy line in Figure 1.1 connects the average monthly temperatures at Oklahoma City over the three-year period 2007–2009. By looking at the heavy line only, it is quite evident that there is a strong annual cycle in temperature. It is equally clear that one sinusoid will not exactly fit all the data, so other harmonics are required. The fundamental period, or period of the first harmonic, is the length of the record, three years. The third harmonic has a period one-third the length of the fundamental period, and consequently represents the annual cycle. The thin line in Figure 1.1 shows the third harmonic after it has been added to the mean of all 36 months, that is, the 0-th harmonic. As expected, the third harmonic provides a close fit to the observed time series.

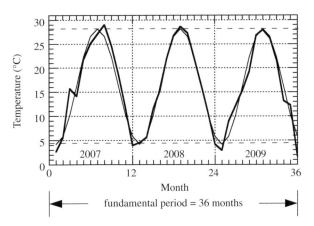

Figure 1.1 Mean monthly temperatures at Oklahoma City 2007–2009 (heavy line), and harmonic 3 (light line) of the Fourier decomposition.

1.1.2 Obtaining the periodogram

The computation of variance arises in elementary statistics as a defined measure of the variability in a data set. When the computation of variance is applied to a time series, it is similarly defined. Now, though, the variance in the data set can be decomposed into individual variances, each one related to the amplitude of a harmonic. Just as adding the sinusoids from all harmonics reproduces the original time series, adding all harmonic variances yields the total variance in the time series. How the decomposition is achieved and how variance is related to harmonic amplitude are discussed in Section 1.2.

A *periodogram* is a plot of the variance associated with each harmonic (usually excluding the 0-th) versus harmonic number and shows the contribution by each harmonic to the total variance in the time series. Henceforth, the term periodogram will be used to refer to the calculation of variance at the harmonic frequencies. The term *Fourier line variance spectrum* is synonymous with periodogram, while the generic term *spectrum* generally means the distribution of some quantity with frequency.

The variance at each harmonic frequency is given by the square of its amplitude divided by two, except at the last harmonic. Figure 1.2 shows the periodogram (truncated to the first 10 harmonics) of the data in Figure 1.1 where we see that harmonic 3 dominates the variability in the data. The small variances at harmonics 6 (period $= 6$ months) and harmonic 9 (period $= 4$ months) are easily observed in Figure 1.2, but, in fact, there are nonzero variances at all 18 possible harmonics (excluding the 0-th) and their sum equals the total variance of $75.23\,°C^2$ in the 2007–2009 Oklahoma City mean monthly temperature time series.

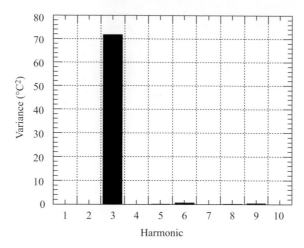

Figure 1.2   Variance at each harmonic through 10 for the data in Figure 1.1.

The periodogram in Figure 1.2 was computed using the computer program given in Appendix 1.A. This program, written in Fortran 77, performs a 'fast' Fourier analysis of any data set with an even number of data and has been used throughout this chapter to compute the periodograms we discuss.

1.1.3 Classification of time series

We can classify time series of data into four distinct types of records. The type of record determines the mathematical procedure to be applied to the data to obtain its spectrum.

The 36 values of temperature x_n, in Figure 1.1, connected by straight-line segments for ease in visualization, constitute a *finite digital* record. Digital time series arise in two ways (Box and Jenkins, 1970, p. 23): sampling an analog time series, for example, measuring continuously changing air temperature each hour on the hour; or accumulating or averaging a variable over a period of time, for example, the previous record of monthly mean temperatures at Oklahoma City. With respect to the latter case, if N is the number of months of data and Δt the time interval between successive values, the record length in Figure 1.1 is $N\Delta t = 36$ months. In this case, as well as with all finite digital records, all data points can be exactly fitted with a finite number of harmonics. This is in contrast to a *finite analog* record of length T, such as a pen trace on an analog strip chart, for example, a seismograph, for which an infinite number of harmonics may be required to fit the signal.

Figure 1.3 is an example of a finite analog record. Sampling the time series at intervals of Δt yields the finite digital record shown in Figure 1.4. The sample values again have been connected by straight-line segments to better visualize the variations

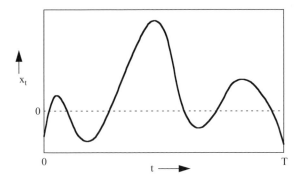

Figure 1.3 An example of a finite analog data record.

in x_n. The sampling interval, Δt, associated with each datum can be shown on a time series plot to the left or right of, or centered on, each datum – it is a matter of choice. In Figure 1.4, Δt is to the right of each datum. One might think that there should be a fifteenth sample point at the very end of the curve in Figure 1.3. However, because of the association of each sampled value with one Δt, the length of the digital record would be one sample interval longer than the analog record. Conceptually, the fifteenth sample point is the first value of a continuing, but unavailable, analog record.

The concept of an *infinite analog* record is often used in theoretical work. An example would be the trace in Figure 1.3 extended indefinitely in both directions of time. For this case a continuum of harmonics is required to fit the signal, thereby resulting in a *variance density spectrum*. Note, however, that a variance density spectrum can be created also with a finite digital record. How this comes about is

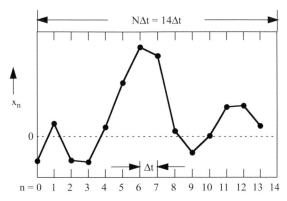

Figure 1.4 An example of a finite digital data record obtained by sampling the finite analog record in Figure 1.3. There are $N = 14$ data.

discussed in Chapter 5. An *infinite digital* record would be obtained by sampling the infinite analog record at intervals of Δt. We will use infinite analog and digital records in Section 3.1.4 (Chapter 3) to determine the effects on the mean value of a time series after it is filtered.

By far the type of record most commonly observed and analyzed in science and technology is the finite digital record. With a few exceptions, this is the type of data record we will deal with in the remainder of Chapter 1, and for which the formulas for computing a periodogram are presented.

1.2 Analysis and synthesis

1.2.1 Formulas

If one of the data sets collected in your research is a time series of atmospheric pressure, Fourier "analysis" can be used to derive its periodogram and to examine which harmonics dominate the series. Conversely, once the analysis has been done, the original time series of pressure can be reconstructed purely from knowledge of the harmonic amplitudes. Thus Fourier "synthesis" is the inverse process of analysis. Note that the title of this chapter employs the more generic meaning of analysis and includes both the analysis and synthesis terms just described.

The formulas in Table 1.1 are those needed to perform analysis and synthesis. The equations under Fourier Analysis are used to calculate the Fourier coefficients or harmonic amplitudes. The equations under Fourier Synthesis express the time series x_n as the sum of products of cosines and sines with amplitudes A_m and B_m, respectively, or, alternatively, the sum of products of cosines only with amplitudes R_m and phase angles θ_m. Notice that the expressions are slightly different depending on whether the time series has an even or an odd number of data. The synthesis equations are equivalent to the forms introduced by Shuster around 1900 (Robinson, 1982).

The arguments of the cosine and sine terms associated with the A_m and B_m coefficients are of the form

$$\frac{2\pi mn\Delta t}{N\Delta t}$$

where m is harmonic number and $n\Delta t$ a point in time along the time axis of total length $N\Delta t$. Thus, $2\pi m$ is the number of radians in the m-th harmonic over the total length of the time series. The product of $2\pi m$ and the ratio $n\Delta t/N\Delta t$ provide location along the sinusoid in radians. Because the time increments (Δt) cancel, they are not shown in Table 1.1. In Fourier synthesis, the summation is over all harmonics at a given location $n\Delta t$, while in Fourier analysis the summation is over all data locations for a given harmonic m.

Table 1.1   Formulas used in Fourier synthesis and analysis for an even or odd number of data.

Fourier Analysis

$$A_0 = \frac{1}{N}\sum_{n=0}^{N-1} x_n \qquad\qquad B_0 = 0$$

$$A_m = \frac{2}{N}\sum_{n=0}^{N-1} x_n \cos\frac{2\pi mn}{N} \qquad\qquad B_m = \frac{2}{N}\sum_{n=0}^{N-1} x_n \sin\frac{2\pi mn}{N}$$

$$m = \left[1, \tfrac{N}{2}-1\right](\text{N even}); \quad m = \left[1, \tfrac{N-1}{2}\right](\text{N odd})$$

$$A_{N/2} = \frac{1}{N}\sum_{n=0}^{N-1} x_n \cos(\pi n) \qquad\qquad B_{N/2} = 0 \quad (\text{N even})$$

$$R_m = \sqrt{A_m^2 + B_m^2} \qquad\qquad \theta_m = \tan^{-1}\left(\frac{B_m}{A_m}\right)$$

Fourier Synthesis

$$x_n = \sum_{m=0}^{N/2}\left(A_m \cos\frac{2\pi mn}{N} + B_m \sin\frac{2\pi mn}{N}\right) = \sum_{m=0}^{N/2} R_m \cos\left(\frac{2\pi mn}{N} - \theta_m\right), \quad n = [0, N-1] \ (\text{N even})$$

$$x_n = \sum_{m=0}^{\frac{N-1}{2}}\left(A_m \cos\frac{2\pi mn}{N} + B_m \sin\frac{2\pi mn}{N}\right) = \sum_{m=0}^{\frac{N-1}{2}} R_m \cos\left(\frac{2\pi mn}{N} - \theta_m\right), \quad n = [0, N-1] \ (\text{N odd})$$

Variance at Harmonic m

$$S_m^2 = \frac{A_m^2 + B_m^2}{2} \qquad\qquad m = \left[1, \tfrac{N}{2}-1\right] \ (\text{N even}); \quad m = \left[1, \tfrac{N-1}{2}\right] \ (\text{N odd})$$

$$S_{N/2}^2 = A_{N/2}^2 \quad (\text{N even})$$

Total Variance

$$S^2 = \sum_{m=1}^{N/2} S_m^2 \quad (\text{N even}) \qquad\qquad S^2 = \sum_{m=1}^{\frac{N-1}{2}} S_m^2 \quad (\text{N odd})$$

The variance at each harmonic for even and odd data lengths is given in Table 1.1 under the heading Variance at Harmonic m. Note that the only exception to the general formula for harmonic variance occurs at $m = N/2$ when N is even. The cosine coefficient at $N/2$ is squared but not divided by two (the sine coefficient is zero). The formulas for the total variance S^2 under the heading Total Variance yield the same variance estimates as the formula

$$S^2 = \frac{1}{N}\sum_{n=0}^{N-1}(x_n - \bar{x})^2 \tag{1.1}$$

for computing total variance directly from the data, in which $\bar{x}$ is the series mean. The two formulas in Table 1.1 are nearly the same, the only difference being that the expression for the upper limit of each summation depends on whether N is even or odd.

1.2.2 Fourier coefficients

The method for obtaining the Fourier coefficients is based on the *orthogonality* of cosine and sine functions at harmonic frequencies, where orthogonality means that the sum of the products of two functions over some interval equals zero. The method entails multiplying both sides of a Fourier synthesis equation by one of the cosine or sine harmonic terms, summing over all n, and solving for the coefficient associated with the harmonic term.

For example, consider multiplying both sides of the first Fourier synthesis equation in Table 1.1 (using the A_m, B_m form) by $\cos\frac{2\pi kn}{N}$ and summing over all n. The second summation on the right-hand side will have the form and result

$$\sum_{n=0}^{N-1} \sin\frac{2\pi mn}{N} \cos\frac{2\pi kn}{N} = 0 \tag{1.2}$$

where m and k are integers. That this sum is zero can be shown with two examples as well as mathematically. The sine and cosine terms for $m=k=1$ are shown in Figure 1.5 and for $m=1$ and $k=2$ in Figure 1.6. The algebraic signs of the sum of cross products within each quadrant are shown at the base of each figure. Because of symmetry, the absolute magnitude of each sum is the same for each quadrant in

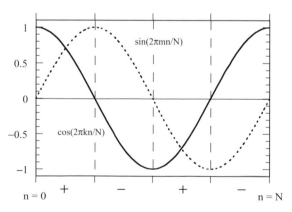

Figure 1.5   Signs of sums of cross products of cosine and sine terms for $m=k=1$.

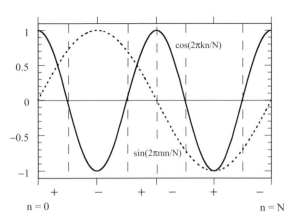

Figure 1.6 Signs of sums of cross products of cosine and sine terms for m = 1 and k = 2.

Figure 1.5 and similarly for Figure 1.6. Thus the waveforms are orthogonal because the sum of their cross products is zero over the interval 0 to N in each illustration.

It can be surmised from these figures that the sum of the cross products is zero over the fundamental period for any combination of the m and k integers. But how could this be shown mathematically? Firstly, we put the sine and cosine terms in complex exponential form, and then expand the summation above using Euler's formula to obtain

$$\sum_{n=0}^{N-1} \sin(2\pi mn/N)\cos(2\pi kn/N)$$

$$= \sum_{n=0}^{N-1} \frac{1}{2i}\left(e^{i2\pi mn/N} - e^{-i2\pi mn/N}\right)\frac{1}{2}\left(e^{i2\pi kn/N} + e^{-2\pi kn/N}\right)$$

$$= \frac{1}{4i}\sum_{n=0}^{N-1}\left(e^{i2\pi(m+k)n/N} + e^{i2\pi(m-k)n/N} - e^{-i2\pi(m-k)n/N} - e^{-i2\pi(m+k)n/N}\right). \qquad (1.3)$$

A procedure is developed in Appendix 1.B for finding the sum of complex exponentials. The final two formulas, Equations 1.B.3 and 1.B.4, are very useful for quickly finding the sums of sines and cosines over any range of their arguments. An example of using the first formula follows.

Consider just the first summation on the right-hand side in Equation 1.3. Let

$$Q = \sum_{n=0}^{N-1} e^{i2\pi(m+k)n/N}. \qquad (1.4)$$

Using Equation 1.B.3, Q becomes

$$Q = \frac{1 - e^{i2\pi(m+k)}}{1 - e^{i2\pi(m+k)/N}}$$

$$= \frac{1 - \cos[2\pi(m+k)] - i\sin[2\pi(m+k)]}{1 - \cos[2\pi(m+k)/N] - i\sin[2\pi(m+k)/N]}$$

$$= 0, \quad m+k \neq 0, N. \tag{1.5}$$

The numerator is zero for all integer values of m and k while the denominator is nonzero except when $(m+k) = 0$ or N, in which cases the denominator is 0 and Equation 1.5 is indeterminate. To evaluate Equation 1.5 for these cases we can apply l'Hopital's rule. The result of taking the first derivative with respect to $(m+k)$ in both the numerator and denominator yields a determinate form with value N. That is

$$Q' = \frac{2\pi\sin[2\pi(m+k)] - i\,2\pi\cos[2\pi(m+k)]}{(2\pi/N)\sin[2\pi(m+k)/N] - i\,(2\pi/N)\cos[2\pi(m+k)/N]}$$

$$= N, \quad m+k = 0, N. \tag{1.6}$$

The same result also can be obtained by substituting 0 or N for $(m+k)$ in Equation 1.4. We observe that the first and fourth summations in Equation 1.3 cancel each other for these values.

We can apply the above procedure to the second term in Equation 1.3. The summation will be zero again, except when $(m - k)$ is 0 or N. Employing l'Hopital's rule yields a determinate form with value N for these cases, similar to Equation 1.6. And again, the same results can be obtained from Equation 1.4. Accordingly, when $(m - k) = 0$ or N, the second and third summations in Equation 1.3 cancel. Thus Equation 1.2 is valid for any integer k or m. This includes the possibility that $(k + m)$ is an integer multiple of N.

Now that we have shown that the summed sine–cosine cross product terms akin to Equation 1.2 must be zero, let us consider the sums of sine–sine and cosine–cosine products resulting from multiplying the first Fourier synthesis equation by $\cos\frac{2\pi kn}{N}$ and summing over all n. Following the procedure in Appendix 1.B we find that

$$\sum_{n=0}^{N-1} \sin\frac{2\pi mn}{N} \sin\frac{2\pi kn}{N} = \begin{cases} 0, & k \neq m \\ \frac{N}{2}, & k = m \neq 0, \frac{N}{2}\,(N\,\text{even}); k = m \neq 0\,(N\,\text{odd}) \\ 0, & k = m = 0, \frac{N}{2}\,(N\,\text{even}); k = m = 0\,(N\,\text{odd}) \end{cases} \tag{1.7}$$

and

$$
\sum_{n=0}^{N-1}\cos\frac{2\pi mn}{N}\cos\frac{2\pi kn}{N} =
\begin{cases}
0, & k\neq m \\[2mm]
\frac{N}{2}, & k = m \neq 0, \frac{N}{2}\,(N\,\text{even}); k = m \neq 0\,(N\,\text{odd}) \\[2mm]
N, & k = m = 0, \frac{N}{2}\,(N\,\text{even}); k = m = 0\,(N\,\text{odd}).
\end{cases}
\tag{1.8}
$$

Thus multiplying the synthesis equation for N even by the k-th sine harmonic term and summing yields

$$
\sum_{n=0}^{N-1}x_n\sin\frac{2\pi kn}{N} = \sum_{m=0}^{N/2}\left(A_m\sum_{n=0}^{N-1}\sin\frac{2\pi kn}{N}\cos\frac{2\pi mn}{N} + B_m\sum_{n=0}^{N-1}\sin\frac{2\pi kn}{N}\sin\frac{2\pi mn}{N}\right)
\tag{1.9}
$$

which reduces to

$$
\sum_{n=0}^{N-1}x_n\sin\frac{2\pi kn}{N} = B_k\,N/2, \quad k = \left[1,\frac{N}{2}-1\right]
\tag{1.10}
$$

so that

$$
B_k = (2/N)\sum_{n=0}^{N-1}x_n\sin\frac{2\pi kn}{N}, \quad k = \left[1,\frac{N}{2}-1\right].
\tag{1.11}
$$

Observe that the sine coefficients for $k = 0$, $N/2$ (N even) are always zero.

The Fourier cosine coefficients, A_k, are obtained in a similar manner, but A_0 and $A_{N/2}$ are, in general, nonzero. As is evident from Table 1.1, A_0 is the mean of the time series. For N odd, an expression similar to Equation 1.9 is used to obtain the Fourier coefficients, the only difference being that the range of harmonics extends from 0 to $(N-1)/2$. Table 1.1 shows the resulting formulas for all Fourier coefficients.

The coefficients A_m and B_m represent the amplitudes of the cosine and sine components, respectively. As shown in the left-hand panel of Figure 1.7a, the cosine coefficient is always along the horizontal axis (positive to the right), and the sine coefficient is always normal to the cosine coefficient (positive upward). In the right-hand panel we see how the cosine and sine vector lengths determine the associated cosine and sine waveforms (ignore the dashed line for the moment). Figures 1.7b–d show various possibilities of waveform relationships depending on the sign of A_m and the sign of B_m. More discussion of Figure 1.7 is given in Section 1.2.4.

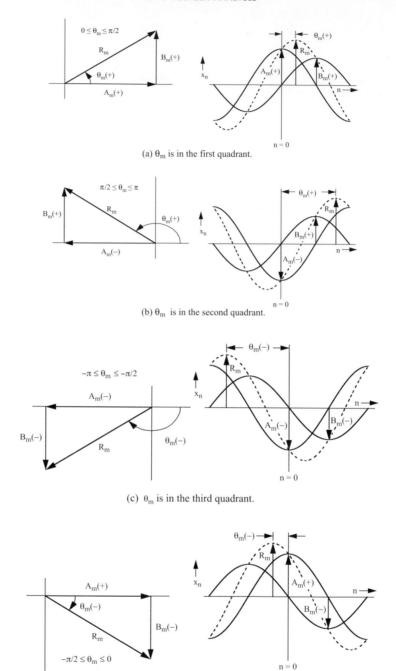

Figure 1.7 (a)–(d) The magnitude and sign of each Fourier coefficient determines the quadrant in which the phase angle lies. Geometric vector lengths in the left hand panels are twice the lengths of the Fourier coefficients in the right hand panels.

An alternative approach can be taken to solve for the Fourier coefficients. As shown by Bloomfield (1976, p. 13), the As and Bs above are identical to the coefficients from a least-squares fit of individual harmonics to the data.

1.2.3 Total and harmonic variances

The standard formula for the total variance of a time series of length N

$$S^2 = \frac{1}{N} \sum_{n=0}^{N-1} (x_n - \bar{x})^2 \tag{1.1}$$

was given in section in Section 1.2.1. The total variance is identical to the sum of the variances at the individual harmonics as shown in Table 1.1 for N even and N odd. The variance at an individual harmonic can be derived from Equation 1.1 by first substituting the Fourier synthesis equations for N even or N odd in Table 1.1 into Equation 1.1 for x_n and $\bar{x}$. The substitution for $\bar{x}$ is A_0. After expanding the synthesis equation inside the parentheses in Equation 1.1, squaring the result, and performing the required summation, the cross product terms vanish (see Equation 1.2) and, using Equations 1.7 and 1.8, the remaining squared terms will reduce to the equations for variance at any harmonic seen in Table 1.1. With one exception, a harmonic variance is the sum of the squares of the Fourier cosine and sine coefficients divided by two. The exception occurs at harmonic $m = N/2$. The expansion of Equation 1.1 into the sum of harmonic variances is a good exercise in the application of orthogonality in time series analysis.

1.2.4 Amplitude and phase representation

Instead of representing a time series by the appropriate sums of **both** sines and cosines, an alternative representation is to use **either** sines or cosines alone and include phase angles, as seen in the right-hand equations in Table 1.1 under Fourier Synthesis. Because of orthogonality, the cosine term is shifted by 90° or $\pi/2$ radians from the sine term for any harmonic. As a result, a single sinusoid can be represented by two amplitude coefficients (A_m and B_m) or, equivalently, by a single amplitude coefficient R_m and a phase angle θ_m. The advantages of the latter are a slightly more compact representation of x_n and only one waveform for each harmonic.

Figure 1.7a illustrates the connection between the two forms of Fourier synthesis. The dashed sinusoid with amplitude R_m in the right-hand panel has been decomposed into a cosine term and a sine term. Their respective amplitudes A_m and B_m depend on the location of the dashed sinusoid relative to the origin $n = 0$, that is, its phase angle θ_m. As noted earlier, the left-hand panel shows the vector relationship among the three amplitudes and the phase angle. Thus

$$R_m = \sqrt{A_m^2 + B_m^2}$$

$$A_m = R_m \cos\theta_m$$

and

$$B_m = R_m \sin\theta_m$$

so that

$$\theta_m = \tan^{-1}(B_m/A_m), \quad -\pi \leq \theta_m \leq \pi.$$

Substituting the middle two equations above into a cosine–sine synthesis results in the amplitude phase synthesis. We see that phase angle θ_m is determined by the sign and magnitude of A_m and B_m. The sign of each coefficient, not merely the sign of the ratio, determines the quadrant in which the phase angle lies. The left-hand panels in Figures 1.7a–d show the amplitude and phase angle in the quadrant associated with the right-hand panels. We observe in each right-hand panel that, given the dashed line and the origin, we can immediately determine the magnitudes of the cosine and sine coefficients: the cosine coefficient is available at the origin and the sine coefficient 90° to the right.

1.3 Example data sets

1.3.1 Terrain heights

Table 1.2 contains the data set for this example. The formulas in Table 1.1 are used to perform a Fourier analysis and synthesis. Consider h to be the variation of terrain height above some datum with distance d along a specified direction. Furthermore, let the data in the table represent a finite digital subset of analog periodic data. The data are plotted in Figure 1.8 and connected by straight-line segments. After looking at Figure 1.8 and the tabled data, it should become clear that the waveform repeats itself every 3000 m. Thus one may as well work with 15 values ($n = 0, \ldots, 14$). Or should one use 16 values? Let us determine the difference. Since $\Delta d = 200$ m and the length of the fundamental period $L = 3000$ m, $N = 15$. Every datum must have a space increment Δd associated with it. Although 16 points subtend L, the Δd associated with the sixteenth point would make the fundamental period 3200, which it clearly is not. In short, the sixteenth point is the first point of the next period and similarly for the thirty-first point in the table.

Table 1.2 Height (h) versus distance (d = nΔd = n200 m).

n	nΔd(m)	h(m)	n	nΔd(m)	h(m)	n	nΔd(m)	h(m)
0	0	100.00	10	2000	98.25	20	4000	101.75
1	200	129.02	11	2200	110.73	21	4200	117.30
2	400	127.14	12	2400	97.55	22	4400	110.59
3	600	102.45	13	2600	72.86	23	4600	89.41
4	800	89.27	14	2800	70.98	24	4800	82.70
5	1000	101.75	15	3000	100.00	25	5000	98.25
6	1200	117.30	16	3200	129.02	26	5200	110.73
7	1400	110.59	17	3400	127.14	27	5400	97.55
8	1600	89.41	18	3600	102.45	28	5600	72.86
9	1800	82.70	19	3800	89.27	29	5800	70.98
						30	6000	100.00

Further examination of the first 3000 m in Figure 1.8 suggests odd symmetry in the data. That is, if a vertical line were drawn at 1500 m, the heights of any two points equidistant from this line will appear to be a reflection about a horizontal line at 100 m elevation. Consequently, except for the mean, only sine terms will be needed in the Fourier synthesis. Lastly, we notice that the time series exhibits only comparatively slow fluctuations, so that most of its variance should be "explained" (i.e., accounted for) by low harmonic frequencies.

Based on this insight, we first compute the mean and find that $A_0 = 100$ m. Over the first 3000 meters we can easily identify three peaks and three troughs, indicating we should calculate the magnitude of harmonic 3, that is

$$B_3 = (2/15) \sum_{n=0}^{N-1} h_n \sin(2\pi 3n/15) = 20 \, \text{m}, \quad N = 15.$$

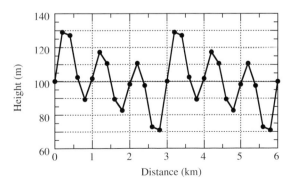

Figure 1.8 Plot of terrain height data connected by straight-line segments.

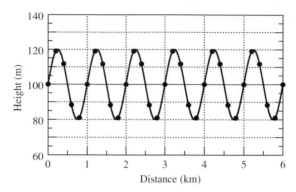

Figure 1.9   Plot of the mean plus harmonic 3 fitted to the data in Figure 1.8 from 0 to 3 km and repeated from 3 to 6 km.

Thus, its variance is $S_3^2 = B_3^2/2 = 200 \text{ m}^2$. The third harmonic added to the mean is plotted in Figure 1.9 from 0 to 3000 m and then repeated to include the entire length of the data set.

Since the average value of the height departures from A_0 for the first 1500 m is positive and for the second 1500 m is the same magnitude but negative in sign, harmonic 1 should be nonzero. This is illustrated in Figure 1.10, in which it can be seen that harmonic 1 will have to be a sine wave to account for heights above the mean from 0 to 1500 m and heights below the mean from 1500 m to 3000 m.

Using the formula for B_1 we get $B_1 = 10$, so that $S_1^2 = 50 \text{ m}^2$. The first harmonic added to the mean is plotted in Figure 1.11, again over the entire time series. The accumulated variance from harmonics 1 and 3 is 250 m^2 in comparison to the total variance of 282 m^2 computed from Equation 1.1. As there is no apparent high frequency variance in Figure 1.8, we would expect much of the remaining variance to be at a low harmonic frequency. If we try harmonic 2 we find that $B_2 = 8 \text{ m}$ and $S_2^2 = 32 \text{ m}^2$. The waveform is shown in Figure 1.12. Since the first three harmonics

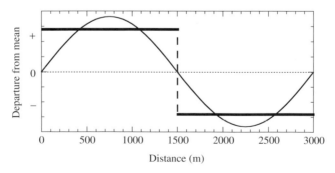

Figure 1.10   Harmonic 1 results from above average and below average heights as shown.

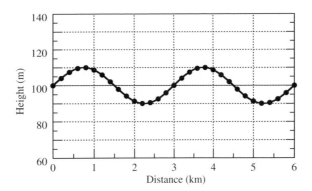

Figure 1.11 Same as Figure 1.9 except for harmonic 1.

account for 100% of the variance in the data, there is no need to make further calculations (obviously, the data were generated using just the above coefficients!). Figure 1.13 shows the sum of the three harmonics plus the mean, drawn as a smooth curve that passes through all the observations in Figure 1.8.

It is interesting to consider what would happen if a 16-point data length (3200 m) were used, an earlier consideration. Instead of computing the three sine coefficients above to explain 100% of the variance, it would take eight cosine and eight sine nonzero coefficients to account for all the variance. The addition of the one point destroys the symmetry present in the 15-point data length (3000 m).

In Figures 1.9–1.13 the waveforms from 0 to 3000 m were repeated over the interval 3000–6000 m. This is allowed since the data are periodic. By extending the Fourier synthesis using the 15-point record, namely

$$h_n = 100 + 10 \sin(2\pi n/15) + 8 \sin(4\pi n/15) + 20 \sin(6\pi n/15)$$

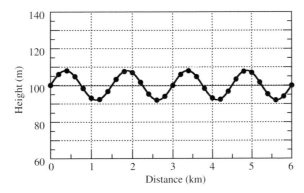

Figure 1.12 Same as Figure 1.9 except for harmonic 2.

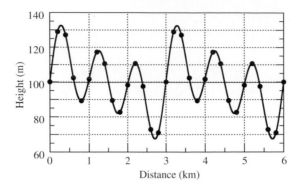

Figure 1.13 Plot of the mean plus harmonics 1, 2, and 3 fitted to the data in Figure 1.8 from 0 to 3 km and repeated from 3 to 6 km.

the series will exactly match that in Figure 1.8 for $15 \leq n \leq 29$, and, more generally, repeat the observed data for any n replaced by $n \pm 15k$ where k is an integer. With a 16-point record, values of the computed series will be repeated every $\pm 16k$ points but will not match the observed record for $k \neq 0$. For example, the first value in the extended series will be 100 m compared to 129.02 m in the observed series.

Clearly, it is important to determine the correct number of data points when working with periodic data. Many observed variables in meteorology and other physical sciences are externally forced by the sun, so that there are strong diurnal and annual components in the data. These components serve to define the fundamental period.

1.3.2 Paradrop days

Table 1.3 shows the mean number of days in January that "paradrop" criteria are met for each hour of the day at Seymour-Johnson Air Force Base, Goldsboro, North Carolina. A paradrop is the insertion of troops or equipment into a site via parachute from an airplane. For a safe paradrop, three meteorological criteria should prevail: ceiling ≥ 2000 feet (610 m), horizontal visibility ≥ 3 miles (4.8 km) and surface winds < 10 knots (5.1 m/s). Although "ceiling" has a specific definition, it can be taken here to mean there is good vertical visibility between the surface and the height of the ceiling. As an example of paradrop days, from Table 1.3 we see that at 0700, 19.2 days of the 31 days in January meet the safety criteria, on average.

The results of performing a Fourier analysis of the data given in Table 1.3 are shown in Table 1.4. Only the results for the three largest harmonics are presented as they account for 97.8% of the total variance and the remaining variances are all small. Figure 1.14 is a plot of the mean number of days of occurrence versus time, the three largest harmonics (about the mean), and their sum. As expected, their sum provides a good fit to the data.

Table 1.3   January paradrop days at Seymour-Johnson Air Force Base, North Carolina.

Hour	Mean number of days	Hour	Mean number of days
0000	21.6	1200	15.6
0100	21.1	1300	15.7
0200	21.2	1400	16.2
0300	20.8	1500	16.5
0400	20.3	1600	18.3
0500	20.4	1700	20.5
0600	20.0	1800	23.0
0700	19.2	1900	23.1
0800	19.5	2000	23.4
0900	18.0	2100	22.4
1000	17.4	2200	22.4
1100	17.5	2300	21.5

We can also establish the time or times of the peaks in each harmonic. To do this, we use the formula

$$H = F \times (\text{phase angle } \theta \text{ in degrees})$$

where H is the time in hours after 0000 local time and F is the ratio of the harmonic period to 360°. Thus, from Table 1.4 for the second harmonic

$$H = (12/360) \times (-143.3) = -4.78 \text{ h.}$$

Converting this value to local time, we get 0713 and 1913. Similarly, the peaks for the third harmonic are at 0223, 1023, and 1823 and for harmonic one at 2305.

In Problem 6 at the end of this chapter you are asked to write a Fourier analysis computer program, apply it to the data in Table 1.3, compare your results with those in Table 1.4 and Figure 1.14, and try to ascribe physical meaning to the main harmonics.

Table 1.4   Results of Fourier analysis of the data in Table 1.3.

Harmonic	Cosine coefficient	Sine coefficient	Variance	Percentage of total variance	Phase angle in degrees
0	19.817	0	0	0	
1	2.8188	−0.6848	4.2072	74.9	−13.7
2	−1.1965	−0.8919	1.1136	19.8	−143.3
3	−0.1743	0.5623	0.1733	3.1	107.2

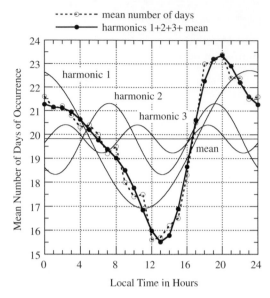

Figure 1.14   Mean number of days that meet paradrop criteria versus time of day at Seymour-Johnson Air Force Base, Goldsboro, North Carolina for the month of January. Three harmonics explain 98% of the variance in the data.

1.3.3 Hourly temperatures

Even if there is a substantial amount of variance at a number of harmonics, one should not believe, in general, that the variance at each harmonic is the consequence of a different physical cause. Instead, often a band of harmonics can be related to a physical phenomenon. Figure 1.15 is an example of a spectrum that shows variance at particular harmonics and at broad bands of frequencies. The data from which the spectrum was computed are hourly temperatures taken at the Norman, Oklahoma Mesonet site (McPherson *et al.* 2007; http://www.mesonet.org) from 1 December 2006 through 31 March 2007 at a height of 1.5 m. Each hourly temperature is a five-minute average at the top of the hour. The logarithmic x-axis is in frequency in cycles/h converted from harmonic number and the y-axis is proportional to the product of variance and frequency. In contrast to the periodogram with line variance in Figure 1.2, the spectrum amplitudes here are connected by straight-line segments, the usual method of presentation. There are two broad frequency bands of interest. One contains periods from about 12 to around 30 days (0.0035–0.0014 cycles/h) and the other from about four to eight days (0.0104–0.0052 cycles/h). The variances in these two bands are due to the passage of long waves in the westerlies (major ridge–trough systems, i.e., Rossby-type waves) and short waves (minor ridge–trough systems and fronts), respectively. Their largely aperiodic nature results in the distribution of variance across a band of

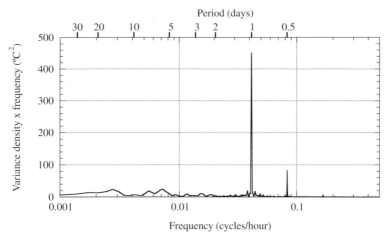

Figure 1.15 Temperature spectrum at 1.5 m height for December 2006–March 2007 at Norman, Oklahoma.

frequencies, the width of which can vary from year to year. Variance at periods longer than 30 days is not shown because there are too few cycles over the 121 days of data to yield satisfactory estimates.

The two well-defined peaks at periods of 24 and 12 hours are due to the daily cycle of solar heating. Similar to the paradrop data in the previous section, the diurnal temperature variation is a deformed sinusoid such that a semi-diurnal component is also required. In fact, close inspection of Figure 1.15 shows a small amount of variance at a period of six hours (0.1667 cycles/h), the fourth harmonic of the daily cycle of temperature. Thus all three variances are required to explain the variance in the daily cycle.

As a final comment about Figure 1.15, we point out that when the product of variance density and frequency is plotted against the logarithm of frequency, the result is an equal-area representation. Thus this is the plot design to use when the goal is to compare variances from different frequency bands. Although we mentioned in Section 1.1.3 that variance density would be discussed in Chapter 5, here it is only necessary to know that variance and variance density are directly proportional to each other to understand Figure 1.15.

1.3.4 Periodogram of a rectangular signal

Fourier analysis necessarily fits sinusoids to a time series; thus it is interesting to observe what happens when data are intrinsically not comprised of sinusoids. The heavy solid line in Figure 1.16a shows a periodic rectangular signal that might represent, for example, whether it is raining or not or the occurrence and non-

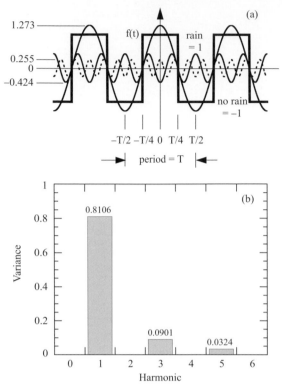

Figure 1.16 (a) A hypothetical rain – no rain analog signal, f(t), showing the first three nonzero cosine harmonics. (b) Resulting periodogram with fundamental period shown in (a).

occurrence of some periodic phenomenon. Because of the location of the time origin, the signal is an even function and, therefore, only Fourier cosine coefficients will be required. By analogy with Fourier analysis for digital data, the expression for the Fourier cosine coefficients for this periodic analog record is

$$A_m = \frac{2}{T} \int_{-T/2}^{T/2} x(t)\cos(2\pi mt/T)dt, \quad m = 0, 1, 2, \ldots$$

where T is the fundamental period. The periodogram in Figure 1.16b shows that only odd-numbered harmonics are needed to synthesize the signal. The two light solid lines and dashed line in Figure 1.16a are the waveforms of the first, third, and fifth harmonics.

Given the waveform of the first harmonic, the waveforms of the third and fifth harmonics serve to improve the Fourier synthesis. The positive and negative over-shoots of the rectangular signal by the first harmonic are compensated by the addition of the associated negative and positive peaks, respectively, in the third harmonic. However, as can be seen in Figure 1.16a, the addition results in overcompensation that, in turn, is compensated by the fifth harmonic. In this way, adding the waveforms of successive odd harmonics better approximates the flat peaks and troughs in the rectangular signal and sharpens the change in value from 1 to -1 and -1 to 1. In practice, if you computed a periodogram that showed evidence of decreasing variance at alternate harmonics, you should be wary of the presence of a rectangular wave. Other special signals (for example, triangular and saw-tooth) also show characteristic spectra.

In summary, we are reminded that Fourier analysis fits sinusoids to data regardless of the signal being generated by the physical (or mathematical) process. It is up to the analyst to keep this in mind when interpreting a periodogram.

1.4 Statistical properties of the periodogram

Section 1.4.1 provides basic statistical concepts and terminology needed to understand the remainder of Section 1.4. Section 1.4.2 discusses the term *expectation* and shows how it is used to find statistical properties, for example mean, variance, and covariance, of digital and analog data. Expectation is used in Appendix 1.C to derive the formulas for the distribution of variances at the Fourier harmonics. Section 1.4.3 deals with the main result of Appendix 1.C, namely, that the frequency distribution of variance at any harmonic is proportional to a chi-square variable. This conclusion requires that the data we analyze come from a normal white noise process. That is, the data have a normal distribution and the periodogram of the data shows no preference for large or small variance at any harmonic. In practice, we assume the data are at least approximately normally distributed and, if the data are not white noise, the data length is sufficiently long that the properties of the chi-square distribution of variance at a harmonic and their independence from one harmonic to the next are effectively met.

Knowing that the statistical distribution of variances at a harmonic is chi-square opens the window to finding confidence limits for the underlying variance spectrum from which a sample periodogram has been computed as well as testing the null hypothesis that the periodogram came from a white noise process. To put these ideas into practice, we will deal with two data sets: one a 100-year record of autumn temperatures; the other a five-year record of monthly mean temperature, both from central Oklahoma, USA. The theory and application of confidence limits are presented in Sections 1.4.4–1.4.6.

1.4.1 Concepts and terminology

The computation of a periodogram is purely an algebraic manipulation of the data. The interpretation of a periodogram depends on how one views the data. If one views a data set as resulting from a physical phenomenon or mathematical process that produces an identical data set each time the phenomenon or process is initiated, the data set is considered *deterministic*. Each data set produced is identical to every other data set, and likewise for the associated periodograms. If one views a data set as resulting from a physical phenomenon or mathematical process that produces a different data set each time the phenomenon or process is initiated, however small the differences, the data set is *nondeterministic*, or equivalently, the data are *stochastic* or *random*. Because no two data sets are alike, there is, conceptually, a *population* of data sets, with each data set a member or *realization* of the population. The population can be finite or infinite. The periodogram of a nondeterministic data set is also one realization of a population of periodograms. In this concept, each realization, whether a data set or a periodogram, represents an equally valid statistical representation of the physical phenomenon or mathematical process being analyzed. Observed time series in natural science are typically nondeterministic, although deterministic components can exist in the series. In Section 1.4 we focus our attention on nondeterministic data sets and the statistical properties of the resulting periodograms. As part of this effort, we need certain additional terminology.

A *random variable* (rv) is a variable that has associated with it a range of values and either a *probability distribution* (pd) if the variable is digital, or a *probability density function* (pdf) if the variable is analog. For example, random variable (rv) X might take on any integer value from 151 to 250, inclusive. The probability distribution gives the probability of occurrence assigned to each of the 100 possible values, with the sum of the probabilities equal to one. Alternatively, rv X might take on any real value within the range 151 to 250, of which there are infinitely many possibilities. In this case the probability of X *exactly* assuming any particular value (say, exactly 200) is zero. However, there exists a finite probability that X will lie within a range of values (say, 199.99–200.01) that is a subset of the overall range of possibilities. Thus, in the case of analog data, it is necessary to describe probabilities using pro-bability densities, the relative likelihoods of the values within the overall range. The probability density function describes these relative likelihoods and, in parallel with the case of digital data, the integral of the probability density function over the range of rv X is one.

Now we develop the concept of a time series of random variables. Imagine a time series of data from time t_0 to time t_T collected from an experiment. Continue to repeat the experiment, thereby forming successive data sets (realizations) of x(t) from t_0 to t_T. The values of x at, say, time t', where $t_0 \leq t' \leq t_T$ form random variable $X(t')$. This concept is illustrated in Figure 1.17, which shows a selection of realizations stacked one upon the other with t_0 and t_T lying beyond the ends of the time axis shown. The intersection of the left-hand vertical dashed line with each realization provides the range of values that comprise rv X at time t'. A random variable also can exist at any other point along the time axis (for example, $X(t'')$ at t''). In the experiment above, the time axis was finite (t_0 to t_T). In general, both the time axis and the number of realizations can be infinite. Whether the time axis $X(t'')$

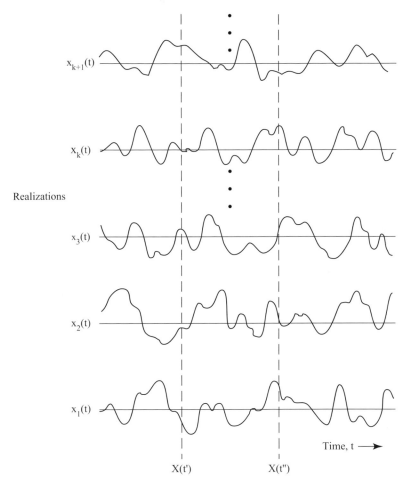

Figure 1.17 A selection of realizations from a random process. Function $X(t')$ denotes random variable X at any time t'. The light horizontal lines have the same reference value of X for each realization.

and number of realizations is finite or infinite t_0 to t_T, the collection of random variables comprise a *random process* or *stochastic process*.

Before continuing with additional concepts we comment on the notation for random variables. Throughout the text we will use only upper case letters to indicate random variables. However, all upper case letters are not random variables; they can be standard mathematical variables, parameters, or constants. We've seen this, for example, with Fourier coefficients in Table 1.1. It is always easy to understand whether or not an upper case letter represents a random variable by the context in which it occurs.

When the data are nondeterministic, we need to consider another attribute: whether the random process is *stationary* or *nonstationary*. If a time series is stationary, the statistical properties of the pd or probability density function do not change with time; a time series is nonstationary if the opposite is true. A simple example of a nonstationary time series is a record of air temperature from winter to summer at a typical middle latitude station. The nonstationarity results from the increasing value of mean daily temperature, that is, a trend. In order to make the series stationary, a low-order polynomial can be fitted to the data set and then subtracted from it. Another common type of nonstationarity occurs when the magnitude of the fluctuations, the population variance, changes with time. An example is the greater variability (gustiness) of the surface wind speed during daytime than nighttime due to the vertical mixing of air as the surface is differentially heated during daylight hours.

If a time series is nonstationary and it is not clear how to remove nonstationary effects, it may be necessary to resort to special analyses using many realizations, divide the data into stationary segments, or apply other methods, such as wavelet analysis (Daubechies, 1992). With the exception of Section 4.1, all mathematical statistics in this text apply to stationary random processes; that is, the population mean and population variance are independent of time. The data sets we analyze can be considered realizations of a stationary random process or can be filtered in such a way to make them stationary or approximately so.

An additional underlying concept needed to derive the statistical properties of a periodogram is a particular random process called *Gaussian* (or *normal*) *white noise*. There are two attributes of this process. With reference to Figure 1.17, the first is that the probability density function of rv $X(t')$ (or $X(t'')$) is Gaussian. The second attribute is that rv $X(t')$ and rv $X(t'')$ are independent of each other. In practice, this means that knowledge of the value of one member of the population at time t' provides no predictability of the value of the same or any other member at any other time. In statistical parlance, the *covariability* or *covariance* between $X(t')$ and $X(t'')$, $t' \neq t''$, is zero, a condition that implies the underlying random process is white noise. The equivalent mathematical statement is derived in the next section.

An examination of the periodogram of any selected time series (a realization) from a Gaussian white noise process would indicate no preference for large or small variances in any part of the spectrum. The average over all possible realizations of the variances at any one harmonic would be identical to that at any other harmonic (with the exception of the highest harmonic for an even number of data, where the variance would be reduced by one-half relative to the other harmonics). That is, the periodogram variances would be uniform with harmonic frequency (less the exception), a condition referred to as "white" by loose analogy with white light wherein no one of the component colors is preferred (there is also acoustic white noise). It is through this connection that the process that produces the uniform variance spectrum is referred to as "white." When we subsequently deal with a white noise process, it will be assumed to be normal or Gaussian, so that use of either qualifier will be dropped.

1.4.2 Expectation

The term *expectation* is of fundamental importance to understanding the statistical properties of the periodogram. The synonym for expectation is average. When expectation is indicated, the average is taken over the entire population, whether it is finite or infinite. The indicator for expectation is the symbol or operator E. When the operator E is applied to a random variable (or a function of a random variable), the question being asked is, "What is the average value of the random variable (or the function of the random variable)." We will see examples of both in subsequent sections.

In the first subsection formulas for the expectation of rv X and general function g(X) for digital data are developed, followed by, in the second subsection, a parallel development for analog data. In the third subsection the expectation of the product of two random variables is developed. The results are formulas for the covariability or covariance between these variables. For those readers familiar with expectation, it may be necessary only to skim through this section to become familiar with the notation.

1.4.2.1 Digital data

Let the sample space S in Figure 1.18 contain a population of N elements, some of which may be the same. Denote distinct elements of S by $x_1, x_2, \ldots, x_K$. In Figure 1.18 $N = 10$ and $K = 6$; two x_1 elements are alike, three x_2 elements are alike, and two x_4 elements are alike.

The symbol for expectation is E and the expectation of random variable X is defined by

$$E[X] = \mu_X = \frac{\sum_{k=1}^{K} x_k n_k}{N} \tag{1.12}$$

where X is a digital random variable, n_k is the number of elements with value x_k, and

$$N = \sum_{k=1}^{K} n_k.$$

The expectation operator asks the question – What is the mean value of the quantity in brackets when the entire population is considered? Thus the summation

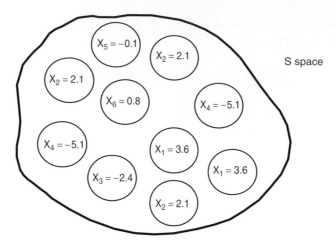

Figure 1.18   Sample space S with N = 10 elements, some of which are alike.

in Equation 1.12 must include all elements. Since the probability of getting value x_k in a random selection of one of the N elements in S is given by

$$P_{x_k} = \frac{n_k}{N}$$

an equation equivalent to Equation 1.12 is

$$E[X] = \mu_X = \sum_{k=1}^{K} x_k P_{x_k}. \qquad (1.13)$$

In the example above

$$N = \sum_{k=1}^{6} n_k = 2 + 3 + 1 + 2 + 1 + 1 = 10$$

and

$$E[X] = \mu_X = \sum_{k=1}^{6} x_k P_{x_k}$$

$$= 3.6 \times \frac{2}{10} + 2.1 \times \frac{3}{10} - 2.4 \times \frac{1}{10} - 5.1 \times \frac{2}{10} - 0.1 \times \frac{1}{10} + 0.8 \times \frac{1}{10}$$

$$= 0.16.$$

Equation 1.13 is good for both finite and infinite populations. In the case of the latter, no empirical determination of the probabilities can be made; they must be known *a priori*.

Now replace digital rv X by a general function of X, namely, g(X). Then,

$$E[g(X)] = \sum_{k=1}^{K} g(x_k) \, p_{x_k}. \tag{1.14}$$

Consider two example functions. Let $g(X) = X^i$ for $i \geq 1$. Then, by analogy with Equation 1.12,

$$E[X^i] = \sum_{k=1}^{K} x_k^i \, p_{x_k} \tag{1.15}$$

is the i-th moment of rv X^i about zero. Of course, $i = 1$ results in the mean μ_x. For the second example function, let $g(X) = (X - E[X])^i$ for $i \geq 1$. The expectation becomes

$$E\left[(X - E[X])^i\right] = \sum_{k=1}^{K} (x_k - \mu_X)^i \, p_{x_k} \tag{1.16}$$

which is the i-th moment about the mean, or the i-th central moment.

A common central moment is the second moment or variance. Accordingly, for $i = 2$ we have

$$E\left[(X - E[X])^2\right] = \sum_{k=1}^{K} (x_k - \mu_X)^2 \, p_{x_k} \tag{1.17}$$

or, what is the same,

$$\mathrm{Var}(X) = \sigma_X^2 = E\left[(X - E[X])^2\right] = E[X^2] - 2E[X] \cdot E[X] + (E[X])^2$$
$$= E[X^2] - (E[X])^2. \tag{1.18}$$

The last form for the variance in Equation 1.18 shows that it is equivalent to the "mean of the squares minus the square of the mean."

For the S space example, substituting Equation 1.13 and Equation 1.15 into Equation 1.18 yields

$$\sigma_X^2 = \frac{[(3.6)^2 \times 2 + (2.1)^2 \times 3 + (-2.4)^2 \times 1 + (-5.1)^2 \times 2 + (-0.1)^2 \times 1 + (0.8)^2 \times 1]}{10} - (0.16)^2$$

$$= 9.732.$$

1.4.2.2 Analog data

The expected value of analog random variable X is given by

$$E[X] = \mu_X = \int_{-\infty}^{+\infty} x\, f(x)\, dx \tag{1.19}$$

where $f(x)$ is the *probability density function* of rv X. Integration is involved for an analog variable as opposed to discrete summation for a digital variable. The limits on X extend over the range $-\infty$ to $+\infty$ and include the case in which the probability density function is zero over some portion of this range. For the general analog function $g(X)$,

$$E[g(X)] = \int_{-\infty}^{+\infty} g(x)\, f(x)\, dx. \tag{1.20}$$

Consider two analog example functions following those for digital data. Let $g(X) = X^i$ for $i \geq 1$. Then

$$E[X^i] = \int_{-\infty}^{+\infty} x^i\, f(x)\, dx \tag{1.21}$$

is the i-th moment of rv X^i about zero. Again, when $i = 1$, we obtain the population mean μ_X. Now let $g(X) = (X - E[X])^i$ for $i \geq 1$. Parallel to Equation 1.16,

$$E\left[(X - E[X])^i\right] = \int_{-\infty}^{+\infty} (x - \mu_X)^i\, f(x)\, dx \tag{1.22}$$

is the i-th central moment. The second moment, $i = 2$, is the variance. Thus, for analog data

$$\mathrm{Var}(X) = \sigma_X^2 = E\left[(X - E[X])^2\right] = \int_{-\infty}^{+\infty} (x - \mu_X)^2\, f(x)\, dx. \tag{1.23}$$

1.4.2.3 Covariance

Consider two analog random variables X_1 and X_2. We can find the variance of each using Equation 1.23. We now inquire about how these two random variables *covary* in time. That is, do they tend to track each other? When X_1 increases, does X_2 also tend to increase (or decrease), or does X_2 just as likely increase as decrease? The measure of this relationship is called *covariance*. If, when X_1 increases (decreases) X_2 also tends to increase (decrease), the sign of the covariance (or covariability) will be positive; if, when X_1 increases (decreases) X_2 tends to decrease (increase), the sign of the covariance (or covariability) will be negative; and, lastly, if, when X_1 increases or decreases X_2 is just as likely to increase as decrease, the expected covariance is zero and the variables are *independent* of each other. Stated mathematically, we have

$$E[(X_1 - \mu_1)(X_2 - \mu_2)] = \mathrm{Cov}[X_1, X_2] = \int_{-\infty}^{+\infty}\int_{-\infty}^{+\infty}(x_1 - \mu_1)(x_2 - \mu_2)f(x_1, x_2)dx_1\,dx_2$$

(1.24)

where $\mathrm{Cov}[X_1, X_2]$ means covariance between random variables X_1 and X_2 and $f(x_1, x_2)$ is the *joint probability density function* between random variables X_1 and X_2.

If X_1 and X_2 are independent, $f(x_1, x_2) = f(x_1) \cdot f(x_2)$; that is, the joint probability density function is equal to the product of the individual probability density functions. With this condition,

$$\mathrm{Cov}[X_1, X_2] = \int_{-\infty}^{+\infty}(x_1 - \mu_1)\,f(x_1)\,dx_1 \int_{-\infty}^{+\infty}(x_2 - \mu_2)\,f(x_2)\,dx_2$$

$$= E[X_1 - \mu_1]\,E[X_2 - \mu_2] = 0.$$

(1.25)

Because the expectation operator is linear, it can be taken inside the brackets of each term in the product on the right, so that $E[X_1 - \mu_1] = \mu_1 - \mu_1 = 0$, and similarly for the second term. The expected value of a constant is, of course, the same constant. The formulas for digital data similar to Equations 1.24 and 1.25 are

$$E[(X_1 - \mu_1)(X_2 - \mu_2)] = \mathrm{Cov}[X_1, X_2] = \sum_{k=1}^{K}\sum_{m=1}^{M}(x_{1k} - \mu_1)(x_{2m} - \mu_2)\,P_{x_{1k}, x_{2m}}$$

(1.26)

where $K = M$ and, for independent variables,

$$\mathrm{Cov}[X_1, X_2] = \sum_{k=1}^{K}(x_{1k} - \mu_1)\,P_{x_{1k}} \sum_{m=1}^{M}(x_{2m} - \mu_2)\,P_{x_{2m}}$$

$$= E[X_1 - \mu_1]\,E[X_2 - \mu_2] = 0.$$

(1.27)

1.4.3 Distribution of variance at a harmonic

Let us now shift our focus from comparing variances among harmonics in Section 1.4.1 to examining how variance is distributed at a single harmonic across the population of realizations. The detailed and somewhat lengthy derivation of this distribution is the subject of Appendix 1.C. In this section we present only the results. Our recommendation is that readers complete this section before studying Appendix 1.C.

Basic knowledge of the properties of a chi-square distribution is essential from this point forward. Sufficient background usually can be found in an undergraduate text in statistics. We will expand on this basic knowledge as needed.

In Appendix 1.C it is shown that, for a normal white noise process, the covariance between the sine and/or cosine coefficients at any two harmonics is zero (a result that might have been anticipated from Equations 1.4 and 1.5) and the coefficients are normally distributed. Squaring the coefficients and standardizing them by dividing by their variances yields random variables with a chi-square distribution. Using the additive property of chi-square variable results in harmonic variances that are independent and proportional to a χ_2^2-distribution (a chi-square distribution with two degrees of freedom) except at the frequency origin (harmonic 0) and, for an even number of data N, at harmonic N/2.

In analyzing geophysical data, we are usually concerned with an underlying stochastic process that is other than white noise. For this situation, the sinusoids at the harmonic frequencies are likewise orthogonal (see Equations 1.4 and 1.5) but it is only in the limit as the number of data N in a realization becomes infinite that their variances are independent and have a chi-square distribution. Koopmans (1974, Section 8.2) provides further discussion of the properties of nonwhite noise.

In obtaining the frequency distribution of variance for a general stochastic process, we will assume N is sufficiently large that it is reasonable to apply the results for white noise given in Appendix 1.C. The magnitude of N required to make this assumption reasonable depends on the departure of the random process from white noise. The greater the departure, the larger the value of N, but no specific value can be given. Thus, in accepting a conclusion from statistical analysis of a realization that depends on it being from a normal white process, it is important to express some caution. Assuming that N is sufficiently large, the variances at the interior harmonics are independently distributed according to

$$\frac{C(f_m)}{\Gamma(f_m)} \Rightarrow \frac{\chi_2^2}{2}, \quad \begin{cases} 0 < m < N/2, & N \text{ even} \\ 0 < m \le (N-1)/2, & N \text{ odd} \end{cases} \qquad (1.28a)$$

where random variable $C(f_m)$ is the variance at the m-th harmonic frequency f_m, $\Gamma(f_m)$ is the process variance at f_m, the arrow indicates "is distributed as," and, therefore, the above variance ratio is distributed as a chi-square variable with two degrees of freedom divided by two. In general, the value of $\Gamma(f_m)$ is unknown. The next section shows how to determine a confidence interval for $\Gamma(f_m)$. The two degrees of freedom (dof) at each harmonic are a consequence of a sine and a cosine being fitted to the data. There is only one dof at the 0-th harmonic (the mean), regardless of whether N is even or odd. In cases where N is even, there also is only one dof at the N/2-th harmonic. As Table 1.1 shows, the calculations at these harmonics require only a cosine term. That is,

$$\frac{C(f_0)}{\Gamma(f_0)} \Rightarrow \chi_1^2, \qquad \text{N even or odd} \tag{1.28b}$$

and

$$\frac{C(f_{N/2})}{\Gamma(f_{N/2})} \Rightarrow \chi_1^2, \qquad \text{N even.} \tag{1.28c}$$

For N odd, the variance at the highest frequency $[(N-1)/2]$ has two dof as noted in Equation 1.28a. For N even or odd the total number of dof in the periodogram equals the number of data N.

It should be noted that $C(f_m)$ is analogous to S_m^2 in Table 1.1. One reason for changing notation is because $C(f_m)$, unlike S_m^2, is a random variable. Another reason is that in Section 1.5.6 we will be calculating variance at any frequency, f, and it will be convenient to simply drop the subscript m. For now, our interest remains in dealing with variance at the harmonic frequencies, f_m, only.

1.4.4 Confidence intervals on periodogram variances

In this and the following section, the underlying stochastic process is unspecified. It may or may not be white noise, but we assume that the number of data, N, is sufficiently large to justify application of independent chi-square distributions derived from a white noise process to the harmonic variances, as discussed in the previous section. In addition, we assume the data follow a normal distribution.

Given that the variance ratio $C(f_m)/\Gamma(f_m)$ follows a chi-square distribution according to Equation 1.28a, we can determine a confidence interval for the ratio using the probability expression

$$\Pr\left\{\frac{\chi_2^2(\alpha/2)}{2} \leq \frac{C(f_m)}{\Gamma(f_m)} \leq \frac{\chi_2^2(1-\alpha/2)}{2}\right\} = 1-\alpha, \qquad m \neq 0 \,(\text{all N}), \, m \neq N/2 \,(\text{N even})$$

$$\tag{1.29}$$

where α is the level of significance. In this equation observed values of $C(f_m)/\Gamma(f_m)$ vary between confidence limits $\chi_2^2(\alpha/2)/2$ and $\chi_2^2(1-\alpha/2)/2$ in $100(1-\alpha)\%$ of the observations. The term $\chi_2^2(\alpha/2)$ is a particular value of the random variable such that the area beneath its probability density function to the left of this value is $\alpha/2$.

We consider the case in which we have an observed value of $C(f_m)$ and the objective is to find the limits of the confidence interval for the population variance $\Gamma(f_m)$. By rearranging Equation 1.29, the $100(1-\alpha)\%$ confidence interval for $\Gamma(f_m)$ can be obtained from the probability statement

$$\Pr\left\{\frac{C(f_m)}{\chi_2^2(1-\alpha/2)/2} \leq \Gamma(f_m) \leq \frac{C(f_m)}{\chi_2^2(\alpha/2)/2}\right\} = 1-\alpha, \quad m \neq 0, N/2 \text{ (N even)}.$$

(1.30)

The interval between $2C(f_m)/\chi_2^2(1-\alpha/2)$ and $2C(f_m)/\chi_2^2(\alpha/2)$ is the $100(1-\alpha)\%$ confidence interval for $\Gamma(f_m)$. By taking the logarithm of the limits of the confidence interval for $\log \Gamma(f_m)$, the lower and upper limits become, respectively,

$$\log C(f_m) + \log(2/\chi_2^2(1-\alpha/2)) \quad \text{and} \quad \log C(f_m) + \log(2/\chi_2^2(\alpha/2)).$$

The logarithmic form of expressing the confidence interval is particularly useful in graphical representations of the periodogram. The reason is that the width of the confidence interval will be fixed regardless of frequency when the variances are plotted on a logarithmic axis.

We now apply Equation 1.30 to a set of data. Figure 1.19 is a plot of the data in Table 1.5 covering 100 consecutive years (1906–2005) of average autumn (September, October, November) temperatures from Climate Division 5 in Oklahoma. Climate Division 5 comprises 13 counties in central Oklahoma. Temperatures are in their original units of degrees Fahrenheit.

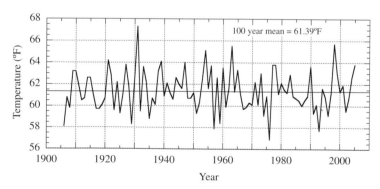

Figure 1.19 One hundred years of mean autumn temperature (September, October, November) for central Oklahoma (Climate District 5) from 1906 to 2005.

Table 1.5   One hundred years (1906–2005) of autumn mean temperature (°F) for Oklahoma Climate Division 5 (central part of the state). (*Source*: Oklahoma Climatological Survey.)

Decade down/ Year across	0	1	2	3	4	5	6	7	8	9
1900–1909							58.1	60.8	59.8	63.2
1910–1919	63.2	61.9	60.5	60.7	62.6	62.6	61.1	59.7	59.7	60.1
1920–1929	60.7	64.2	62.9	59.6	62.2	59.3	61.1	63.8	61.8	58.3
1930–1939	62.4	67.3	59.5	63.6	62.2	58.8	60.7	60.1	63.2	64.1
1940–1949	60.9	62.1	61.2	60.6	62.6	62.0	61.6	64.0	60.7	60.7
1950–1959	61.2	59.3	60.3	62.5	65.1	61.6	63.7	57.9	62.6	58.4
1960–1969	63.5	59.9	61.6	65.5	61.3	63.3	61.1	59.7	59.9	60.3
1970–1979	60.1	62.2	60.1	63.0	59.1	60.9	56.9	63.8	63.8	61.1
1980–1989	62.1	61.5	61.2	62.9	60.9	60.7	60.5	60.0	60.5	60.9
1990–1999	63.6	59.3	60.1	57.7	61.6	60.8	59.1	61.2	65.7	63.0
2000–2009	61.3	61.9	59.5	60.7	62.6	63.8				

The periodogram is shown by the solid line in Figure 1.20 and was computed using subroutine Foranx in Appendix 1.A. The 95% confidence interval for the population variance is shown on the right-hand side of the figure (solid line) where the dot is to be placed over each sample variance $c(f_m)$ as shown, for example, at harmonic 39. Note that $c(f_m)$ is used to denote a sample value of the rv $C(f_m)$. That the dot with constant width confidence interval around it may be placed at any harmonic is a direct consequence of the logarithmic plot, as described above. Because the periodogram varies wildly, it is not easy to discern bands of small or large variance or a trend in variance with harmonic number. Correspondingly, the 95% confidence interval ($\alpha = 0.05$) for $\Gamma(f_m)$ is very wide. The variability in $c(f_m)$ seen here is typical of periodograms of many kinds of geophysical data and is the chief reason that periodograms of observed data are often smoothed, as discussed in the next section.

1.4.5 The smoothed periodogram

To better distinguish bands of large and small variance or a trend in the spectrum, a common practice is to smooth the spectrum by weighting together a number of contiguous variances. The simplest smoothing is the running mean of length n (odd) given by

$$\overline{C}(f_m) = \frac{1}{n} \sum_{j=m-(n-1)/2}^{m+(n-1)/2} C(f_j) \tag{1.31}$$

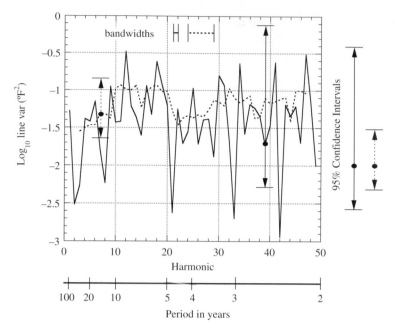

Figure 1.20   Periodogram of the data in Table 1.5 and Figure 1.19 (solid line). Averaged periodogram using 5-point running mean (dashed line). The respective 95% confidence intervals for the population mean variance at each harmonic are shown to the right and the respective bandwidths at the top of the figure.

in which average variance is calculated only for those harmonics that do not include f_0 and $f_{N/2}$ (N even) in the summation. Random variable C is used in Equation 1.31 to indicate we are determining the effects of smoothing on the distribution of variances; in application, however, harmonic variances from a single realization would be smoothed, and lower case variable c would be used as in the previous section. Because of the inability to include the correct number of terms, there will be a loss of $(n-1)/2$ harmonic variances at either end of the smoothed periodogram.

If we assume, as prescribed earlier, that the number of data in a realization is sufficiently large that the variance ratios $C(f_j)/\Gamma(f_j)$ can be approximated by independent χ^2 variables with two dof divided by two and, furthermore, that $\Gamma(f_j)$ is effectively constant over the length n, then, using Equation 1.31, the smoothed variance ratios $\overline{C}(f_m)/\overline{\Gamma}(f_m)$ are approximately χ^2 random variables with 2n dof (Hoel, 1962, p. 268) divided by 2n and are independent every n harmonics.

The dashed line in Figure 1.20 is the result of a five-point running mean and provides an improved picture of the structure of the variance. There are now $2n = 10$ dof associated with each variance ratio. For this realization, periods from 5 to 10 years contain more variance than periods shorter than five years except near the two-year period. This comparison suggests the data should comprise sharp year-to-year fluctuations superimposed on long-period fluctuations. The plot of the data in Figure 1.19 clearly indicates that there are short period variations; long period variations, though, are less obvious. Thus the need for a periodogram to show the not-so-obvious.

By analogy with the limits of the confidence interval for $\log \Gamma(f_m)$, the limits for $\log \overline{\Gamma}(f_m)$ are

$$\log \overline{C}(f_m) + \log(2n/\chi^2_{2n}(1 - \alpha/2)) \quad \text{and} \quad \log \overline{C}(f_m) + \log(2n/\chi^2_{2n}(\alpha/2)).$$

For $n = 5$ and $\alpha = 0.05$, the values for the constant terms above are -0.31 and 0.49. The 95% confidence interval is shown by the dashed vertical line on the right of Figure 1.20 and its reduced length relative to that for no smoothing ($n = 1$) reflects its application to an averaged spectrum, namely, to $\log \overline{\Gamma}(f_m)$, which, by our previous assumption, is approximately $\log \Gamma(f_m)$.

To take into account smoothing of the periodogram by other than a running mean, an approximate general formula for the dof r in χ^2 distributions is, for n odd,

$$r = 2 \left[\sum_{j=-(n-1)/2}^{(n-1)/2} K^2(f_j) \right]^{-1} \tag{1.32}$$

where $K(f_j)$ is a symmetric weight function centered at frequency f_0 such that the sum of the weights is unity (Koopmans, 1974, p. 273). Maintaining unity preserves the total variance in the spectrum. When $K(f_j) = 1/n$, the running mean, $r = 2n$.

Associated with dof is bandwidth β, the frequency interval between independent adjacent estimates of variance. In the case of a periodogram with no smoothing it is

$$\beta = \frac{1}{N\Delta t} \tag{1.33}$$

which is the frequency difference between harmonics i and $i + 1$. Rewriting Equation 1.33 in the form

$$\beta N \Delta t = 1 \tag{1.34}$$

we see that the product of β and $N\Delta t$ is constant. This means that as the length N of a time series increases (Δt remains fixed), the bandwidth of each independent periodogram estimate will proportionately decrease and the total number of spectrum estimates will proportionately increase. Because there are two dof associated with each spectrum estimate, increasing the length of a time series in and of itself does not reduce the variability of periodogram estimates.

Equation 1.33 is exact for white noise and approximate for nonwhite noise when N is large. For a five-point running mean the bandwidth would be five times as wide. The bandwidths are shown in Figure 1.20 for their associated spectra. An approximate general formula for the bandwidth is (Koopmans, 1974, p. 277)

$$\beta = \frac{r}{2N\Delta t} = \left[N\Delta t \sum_{j=-(n-1)/2}^{(n-1)/2} K^2(f_j) \right]^{-1} \qquad (1.35)$$

which reduces to $\beta = n/N\Delta t$ for a running mean of length n.

As an example of a simple nonrunning mean filter, consider a three-point smoother (a triangular filter) whose weights are $\frac{1}{4}$, $\frac{1}{2}$, $\frac{1}{4}$ (sum of weights $= 1$). From Equation 1.32 the dof will be $5\frac{1}{3}$ whereas the number of dof for a three-point running mean is six. From Equation 1.35 the bandwidth of the former is 8/9 as wide as that of the latter. It is of interest to know that the periodogram used to produce the spectrum of hourly temperatures in Figure 1.15 was smoothed with this triangular filter prior to creating the product of variance density and frequency. The purpose was to magnify the two broad frequency bands that were discussed relative to the main peak of the daily cycle of temperature.

1.4.6 Testing the white noise null hypothesis

In this section we examine the problem of testing the null hypothesis that a sample of data comes from a random process that is white noise. This is equivalent to the null hypothesis that the expected values of the spectrum variances are uniform with frequency. A white noise test can be an important tool in analyzing spectra of geophysical data. If we observe in a given spectrum a single variance, multiple variances, or a band of variance that seems to be unexpectedly large, the question arises whether these features are a consequence of an underlying physical process or whether they occurred by chance. If the white noise null hypothesis applied to the spectrum cannot be rejected, there is then doubt that the observed large variance or variances are anything more than natural fluctuations in a realization from a white noise process.

In the previous section a method was developed to place a confidence interval for the population variance surrounding each sample variance (variance at a harmonic from a single realization). In contrast, in this section we will place confidence intervals for the sample variances about the estimated population variance, hypothesized to be uniform with frequency. Here, however, it is necessary to consider two types of confidence intervals. These will be demonstrated with two examples, the first of which employs the 100-year record of central Oklahoma temperature data seen in the previous two sections. How these confidence intervals are used in making a white noise test requires some background knowledge, to which we now turn.

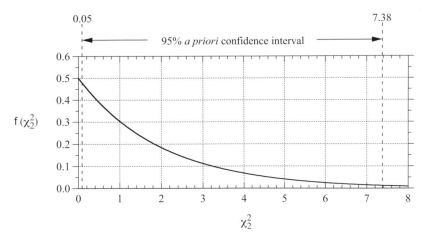

Figure 1.21 The probability density function (pdf) of a random variable that has a chi-square distribution with two degrees of freedom. Confidence limits for the 95% *a priori* confidence interval are shown by the vertical dashed lines. The area under the curve has unit value.

Figure 1.21 shows the probability density function (or, equivalently, the frequency distribution) of a χ^2 random variable with two dof. The probability density function is given by:

$$f(\chi^2_2) = \frac{\exp\left(-\dfrac{\chi^2_2}{2}\right)}{2}. \tag{1.36}$$

The two vertical dashed lines encompass what is called the *a priori* 95% confidence interval. Figure 1.21 is the typical way of presenting a probability density function or frequency distribution with confidence limits. When a confidence interval is applied to a spectrum, the width of this interval is oriented in the vertical, as we did in Figure 1.20.

Let us imagine successively withdrawing 29 samples from a population that has the distribution shown in Figure 1.21. Prior to the first withdrawal, the probability that its value will lie outside the interval (0.05, 7.38) is 0.05. Prior to the second withdrawal, the probability that its value will lie outside the same interval is also 0.05. Repeat this 27 more times. Because each withdrawal is independent of any other, the probability is 0.05 that any sample value of χ^2_2 will lie outside the interval (0.05, 7.38).

Now arrange the sample values of χ^2_2 as shown in Figure 1.22, except that each value withdrawn is divided by two (with this adjustment we can apply the results of this section directly to the distribution of periodogram variance ratios). Before even looking at the sample values of $\chi^2_2/2$, we would not be surprised to find one or two lying outside the confidence interval. This follows from the calculation

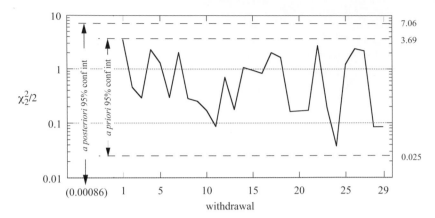

Figure 1.22 A plot of 29 random withdrawals from a chi-square distribution with 2 degrees of freedom after dividing the value of each withdrawal by 2. The chi-square distribution function is shown in Figure 1.21. 95% *a priori* and *a posteriori* confidence intervals are also shown.

$29 \times 0.05 = 1.45$, where 29 is the number of withdrawals and 0.05 is the level of significance or the probability of rv $\chi_2^2/2$ being greater than 3.69 (7.38/2 from Figure 1.21) or less than 0.025 (0.05/2) per withdrawal. (If we had 100 such data sets we would expect 145 of the 2900 values to lie outside the confidence interval.) In fact, Figure 1.22 shows that one value or point lies very close to the upper confidence limit (withdrawal 1 is 3.501) and the value of withdrawal 24 (0.038) is slightly above the lower confidence limit.

To show the probability, α, of observing one or more values from a $\chi_2^2/2$ distribution outside the *a priori* confidence interval, we make use of the binomial distribution

$$\frac{M!}{Z!(M-Z)!} \times p^Z(1-p)^{M-Z}$$

where: p = probability that the value of a randomly selected point will lie outside the *a priori* confidence interval (0.05 for a 95% confidence interval); M = total number of points (29 in this example); and Z = the number of the M points that lie outside the *a priori* confidence limits.

The probability of one or more points lying outside the confidence interval is one minus the probability of no points lying outside the confidence interval, or, in general, $\alpha = 1 - (1-p)^M$. Thus, if $p = 0.05$ and $M = 29$, then $\alpha = 1 - (0.95)^{29} = 1 - 0.2259 = 0.7741$. Instead of having a 5% chance of finding at least one value of $\chi_2^2/2$ outside the confidence interval, we actually have a 77% chance when considering the group of 29 values or points.

In practice, we are sometimes faced with the following dilemmas. In a given data set the number of values that lie outside the *a priori* confidence interval is about as expected, but one of the values is very large. Is the very large value significantly greater than expected? In another case, a few more values than expected lie outside the confidence interval. Is the difference between the expected number and observed number significant?

The solution to both dilemmas is as follows. We really want α to be 0.05. That is, when we consider all the points in the group (29 in this example), we want to find the two particular values of $\chi_2^2/2$ such that there is only a 5% chance that *any one or more* points will lie outside the associated interval. This is called the *a posteriori* confidence interval. With the meaning of p the same as that given previously, except that it now applies to the confidence interval for the group of points, the determination of the limits of this interval follows.

From above, and using the binomial theorem, for $p \ll 1$

$$\alpha = Mp,$$

so that

$$p = \alpha/M.$$

With

$$\alpha = 0.05, \quad M = 29,$$

then

$$p = 0.00172.$$

If we now integrate the probability density function (Equation 1.36) between 0 and χ_2^{2*} and between χ_2^{2**} and ∞ where * and ** indicate particular values of χ_2^2, and equate both results to p/2, we obtain $\chi_2^{2*}/2 = 0.00086$ and $\chi_2^{2**}/2 = 7.06$. These are the lower and upper limits for the *a posteriori* 95% confidence interval and are plotted in Figure 1.22 (however, the lower limit is off the graph). The *a posteriori* confidence interval deals with all 29 values at one time and the *a priori* confidence interval deals with one value at a time. There is only one chance in 20 that any one or more of the 29 values would lie outside the 95% *a posteriori* confidence interval, and as Figure 1.22 shows none do. This result is in accord with our withdrawal of 29 random samples from a χ^2 distribution with two dof divided by two.

In periodogram analysis we typically use *a posteriori* confidence limits because we want to observe the *entire spectrum* of harmonic variances *after the fact* of calculating the spectrum. If we wanted to know whether the variance at a *particular harmonic* exceeded the confidence limits *before the fact* of observing the variance at the harmonic in question, we would use the *a priori* confidence interval. Interest in the latter approach is uncommon. Nevertheless, obtaining *a priori* confidence limits is always a natural first step because if none of the spectrum variances exceed these limits, there is no need to proceed to the next step of computing *a posteriori* confidence limits.

To better understand white noise testing we examine two applications to real data.

1.4.6.1 *White noise test: Example 1*

We revisit the 100-year record of mean autumn temperatures for central Oklahoma (Climate Division 5) given in Table 1.5 and plotted in Figure 1.19. Figure 1.20 showed confidence intervals for the population variance $\Gamma(f_m)$ and smoothed population variance $\overline{\Gamma}(f_m)$ at each harmonic given samples of $C(f_m)$ and $\overline{C}(f_m)$, respectively. The underlying random process was unspecified, but the number of data was assumed sufficiently large to justify using independent chi-square distributions of the harmonic variances derived for a normal white noise process. In this example we will use the same data to find a different kind of confidence interval; that is, we will find confidence limits for observations of rv $C(f_m)$ common to all harmonic frequencies given an estimate $\hat{\Gamma}(f_m)$ of the population variance under the white noise hypothesis. Because a white noise process is hypothesized, there is no restriction on the size of a data set.

The variance of the data set in Table 1.5 is $3.4201°F^2$. As a consequence, our estimate of the population variance at each of the 49 interior harmonics ($m = 0$ and $m = 50$ excluded) in the periodogram under the white noise hypothesis is $\hat{\Gamma}(f_m) = 3.4201°F^2/49.5 = 0.0691°F^2$. The symbol ^ means "estimate of" and the reason we make this distinction is that variance of the realization ($3.4201°F^2$) is just an estimate of the population variance. In general, we do not test the variance at the highest harmonic for N even, here $m = 50$, because its variance, under a white noise hypothesis, is one-half the interior variances; it is a unique harmonic. The reason for its uniqueness is that its bandwidth is one-half the bandwidth associated with each of the interior harmonics. That the divisor is 49.5 instead of 49 is because the variance of the data set included the variance at $m = 50$. Thus, for the general case of an even number of data, N, the white noise variance at the interior harmonics is determined from the total variance in the data set divided by $(N/2 - 1) + 0.5 = (N/2) - 0.5$. In the general case of an odd number of data, N, the white noise variance at all the harmonics (except $m = 0$) is the total variance in the data set divided by $(N - 1)/2$. The highest harmonic has full bandwidth.

To find the *a priori* confidence interval for the sample variances about their estimated expected value, we rewrite Equation 1.29 to obtain the form

$$\Pr\left\{\frac{\Gamma(f_m)\,\chi_2^2(\alpha/2)}{2} \le C(f_m) \le \frac{\Gamma(f_m)\,\chi_2^2(1-\alpha/2)}{2}\right\}$$

$$= 1-\alpha, \quad m \ne 0 \text{ (all N)}; \quad m \ne N/2 \text{ (N even).} \tag{1.37}$$

To be consistent with Figure 1.20, we take logarithms of the limits of the confidence interval and obtain

$$\log\left[\Gamma(f_m)\,\chi_2^2(\alpha/2)/2\right] \quad \text{and} \quad \log\left[\Gamma(f_m)\,\chi_2^2(1-\alpha/2)/2\right]$$

which, for $\alpha = 0.05$ and $\hat{\Gamma}(f_m) = 0.0691°F^2$, are -2.76 and -0.59, respectively. We expect, on average, 2½ (0.05×50) variances to exceed these limits; in Figure 1.23 we note that the variances at three harmonics (12, 42, and 47) fall outside either the upper or lower limit. Should the white noise null hypothesis be rejected? This is a good example in which the answer can be found by calculating the *a posteriori* confidence limits.

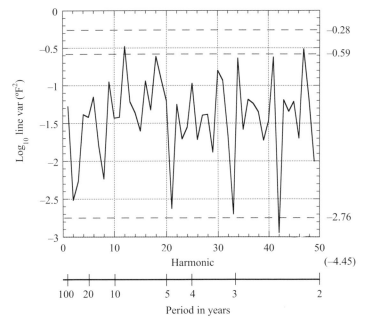

Figure 1.23 Periodogram of the data in Table 1.5 and Figure 1.19 (see also Figure 1.20). The inner two dashed lines are the 95% *a priori* confidence limits; the upper dashed line is the upper 95% *a posteriori* confidence limit. The lower 95% *a posteriori* confidence limit is located below the graph and has value -4.45.

The *a posteriori* confidence interval is determined by replacing $\alpha/2$ in Equation 1.37 and the expressions for the limits of the confidence interval by $p/2$ where $p \approx \alpha/M$ and $M = 49$. Therefore, the parallel equations for *a posteriori* confidence limits are

$$\Pr\left\{ \frac{\Gamma(f_m)\, \chi_2^2(p/2)}{2} \leq C(f_m) \leq \frac{\Gamma(f_m)\, \chi_2^2(1-p/2)}{2} \right\}$$

$$= 1 - p, \quad m \neq 0 \text{ (all N)}; \; m \neq N/2 \text{ (N even)} \tag{1.38}$$

and logarithms of the limits of the confidence interval are

$$\log\left[\Gamma(f_m)\, \chi_2^2(p/2)/2\right] \quad \text{and} \quad \log\left[\Gamma(f_m)\, \chi_2^2(1-p/2)/2\right].$$

Equation 1.36 can be integrated, as in the previous example, to obtain the values of χ_2^2 for area $p/2 = 0.0005102$ at the left and right extremes of the chi-square distribution (refer to Figure 1.21). The results are $\chi_2^2(p/2) = 0.0010207$ and $\chi_2^2(1-p/2) = 15.161$, so that the logarithms of the lower and upper limits of the 95% *a posteriori* confidence interval are -4.45 and -0.28. We observe in Figure 1.23 that no variance lies outside this range and, therefore, we cannot reject the null hypothesis that the data are a realization from a white noise process. Thus there appears to be no useful statistical predictability of mean autumn temperature at Oklahoma City other than using its long-term mean as the predictor.

1.4.6.2 White noise test: Example 2

For our second example, we investigate five years of mean monthly temperature at Oklahoma City from 2003–2007. The data are given in Table 1.6 and plotted in Figure 1.24a. As expected, the time series shows a strong annual solar influence. We can consider the annual solar cycle for each year to vary in a different way about a long-term mean annual temperature cycle. Ideally, it is the long-term cycle we would like to remove from the time series before we apply a white noise test. The best we can do, however, is to estimate this cycle using the five years of data available to us. Except for the solar cycle, the approach to obtain confidence limits is similar to that in the first example.

We can estimate the long-term annual cycle by averaging the five years of data month-by-month and then subtracting the appropriate five-year average from each observed monthly mean. The set of residuals, also given in Table 1.6, form a sequence of 60 values from January 2003 through December 2007 and comprise the time series for which the white noise null hypothesis will be tested. The time series is shown in Figure 1.24b.

Table 1.6 (a) Monthly mean temperatures (°C) at Oklahoma City Will Rogers Airport from 2003 to 2007. The bottom row shows monthly means averaged over the five-year period. (b) Monthly mean residuals, i.e., the appropriate five-year monthly average has been subtracted from each monthly mean. (Note: All monthly means in (a) have been converted to Celsius from the original monthly means in Fahrenheit. *Source*: National Climatic Data Center, Asheville, NC.)

(a)

Month/ Year	Jan	Feb	Mar	Apr	May	Jun	Jul	Aug	Sep	Oct	Nov	Dec
2003	2.67	3.17	9.72	15.78	20.61	23.28	29.06	28.22	20.78	17.56	10.28	6.17
2004	4.39	4.39	12.94	16.22	22.17	24.11	26.06	24.78	23.94	18.11	10.33	6.39
2005	4.22	8.22	10.94	16.28	20.67	25.56	26.89	27.22	25.06	17.50	12.11	3.83
2005	8.72	5.39	12.83	19.61	22.56	26.67	30.11	29.94	21.83	17.11	11.61	6.39
2007	2.67	5.61	15.67	14.11	21.67	25.06	27.06	29.00	24.50	18.61	11.61	3.94
Mean	4.53	5.36	12.42	16.40	21.53	24.93	27.83	27.83	23.22	17.78	11.19	5.34

(b)

Month/ Year	Jan	Feb	Mar	Apr	May	Jun	Jul	Aug	Sep	Oct	Nov	Dec
2003	−1.87	−2.19	−2.70	−0.62	−0.92	−1.66	1.22	0.39	−2.44	−0.22	−0.91	0.82
2004	−0.14	−0.97	0.52	−0.18	0.63	−0.82	−1.78	−3.06	0.72	0.33	−0.86	1.04
2005	−0.31	2.87	−1.48	−0.12	−0.87	0.62	−0.94	−0.61	1.83	−0.28	0.92	−1.51
2006	4.19	0.03	0.41	3.21	1.02	1.73	2.28	2.11	−1.39	−0.67	0.42	1.04
2007	−1.87	0.26	3.24	−2.29	0.13	0.12	−0.78	1.17	1.28	0.83	0.42	−1.40

Before applying the test, a few comments are in order concerning the method of removing the annual cycle. The total variance of the annual cycle is the sum of variances from harmonics with periods of 12, 6, 4, 3, 2.4, and 2 months. When the periodogram of the residuals is computed (an exercise we recommend), one will discover the variance is zero at these six harmonics. The reason is that we have removed the variances at all harmonics of the annual cycle from the original time series. In fact, had we computed a periodogram of the original data, it would have included the identical variances at the harmonics corresponding to the periods of the estimated annual cycle. In this example, we will replace the zero variances at periods of 12, 6, 4, 3, and 2.4 months by the average of the two adjacent variances. While these replacement values are artificial and are not part of the white noise test, for the sake of appearance they will provide a smoothly varying periodogram in the vicinity of the harmonics of the annual cycle. From Equation 1.28c, the distribution of the variance ratio at the harmonic corresponding to a period of two months is χ_1^2. As in Example 1, the white noise test is applied only to the interior harmonics.

The variance of the residual data shown in Figure 1.24b is $2.3160\,°C^2$. Under the null hypothesis that this data set is a realization of a white noise process, we can

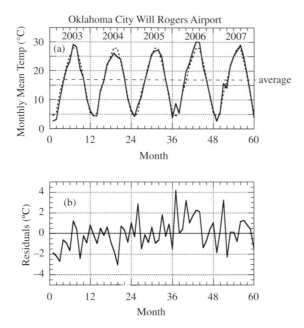

Figure 1.24 (a) Mean monthly temperatures at Oklahoma City Will Rogers Airport from 2003 to 2007 (solid line) and average mean monthly temperatures (dashed line). (b) Residual mean monthly temperatures (actual − average).

estimate the population variance $\Gamma(f_m)$ at each of the 29 interior harmonics (harmonics $m = 0$ and $m = N/2$ are excluded) in the Fourier spectrum using the equation $\hat{\Gamma}(f_m) = (2.3160/24)\,°C^2 = 0.0965\,°C^2$. The reason for dividing by 24 is that the residual variance does not include any variance from the six harmonic frequencies previously discussed. Since there are a total of 30 harmonic frequencies (60 samples of data), and the variance has been removed from the six harmonics associated with the annual cycle, the residual variance is distributed equally among the remaining 24 harmonics to estimate the population variance.

Upon replacing $\Gamma(f_m)$ by its estimate $\hat{\Gamma}(f_m)$, we conclude from Equation 1.28a that the variance ratio $C(f_m)/\hat{\Gamma}(f_m)$ varies approximately as $\chi_2^2/2$. Figure 1.25 shows the sample variance ratios versus harmonic number where the ratios at harmonics 5, 10, 15, 20, and 25 are the averages of adjacent ratios. Since no ratio lies outside the *a posteriori* confidence interval, the null hypothesis that the sample data come from a random process that is white noise cannot be rejected at the 5% level of significance. Stated another way, this realization can be viewed as a member of a population of similar random time series, the totality of which comprises a white noise random process. In this particular example, computing the *a posteriori* confidence interval was not necessary since none of the variance ratios lie outside the *a priori* confidence interval. The goal of this example was to derive the 95% *a posteriori* confidence limits for variance ratios as opposed to variances in the first example.

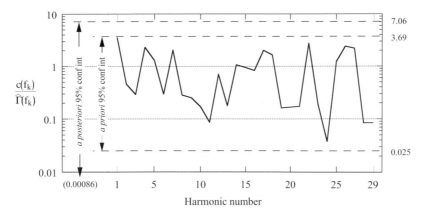

Figure 1.25 Observed variance ratio versus harmonic frequency for the residuals in Figure 1.24b. The population variance $\Gamma(f_m)$ is estimated from the sample variance. Harmonic frequency f_m has been converted to harmonic number. 95% *a priori* and *a posteriori* confidence intervals are also shown.

A keen observer will recognize that the plot in Figure 1.22 is identical to that in Figure 1.25. In fact, the same data set was used to produce the plot in Figure 1.22. Thus withdrawing 29 values from a chi-square distribution in the discussion in Section 1.4.6 was a little "white" lie! But whether one literally withdrew samples from a chi-square distribution was immaterial to developing an understanding of the mechanics of a white noise test.

Practically speaking, if the data set in this example is representative of other five-year intervals at Oklahoma City, then there is no skill in attempting to forecast mean monthly temperature beyond what can be accomplished by employing the average annual cycle. Had the white noise hypothesis been rejected, there would have been potentially useful skill in mean monthly temperature forecasts apart from the average annual cycle. In conclusion, if there is interest on the part of an investigator to make statistical forecasts of any variable represented by a time series, a good first step is to perform a white noise test of the original data or, if appropriate, the residual data, that is, the original data less the deterministic components.

1.5 Further important topics in Fourier analysis

At this juncture, we are able to (i) compute the Fourier coefficients of a data set, (ii) calculate its spectrum or periodogram, (iii) determine a confidence interval for the population variance at each harmonic frequency, and (iv) perform *a priori* and *a posteriori* white noise tests. Now we consider selected topics that will extend our understanding of Fourier analysis. As we have already seen in Table 1.1, the number of harmonics at which variance is computed is directly related to the number of data.

Section 1.5.1 explains why. The second topic, covered in Section 1.5.2, shows, mathematically, why variance calculated at a given harmonic frequency includes not only the variance at that harmonic but also variance from frequencies between nearby harmonics. Thus variance in one part of a spectrum can "bleed" or "leak" to another part of the spectrum. In short, we always view a spectrum through a "window."

Sometimes we are faced with a signal and noise problem. For example, we might be suspicious that there is a 60 Hz signal, say, from a power source, corrupting a data set we collected. That is, a deterministic signal may be embedded in otherwise random data. In Section 1.5.3 we investigate how averaging spectra from a number of data records, each of which contains the deterministic signal, smooths the averaged spectrum so the deterministic signal is more easily discernible. Another approach is discussed in Section 1.5.4, where we examine the effect that increasing the length of a time series has on discriminating a sinusoid from random components. The fifth topic shows how to convert the formulas in Table 1.1 for Fourier synthesis and analysis to complex form; this is developed in Section 1.5.5. Because of trigonometric symmetry, a complex representation is very compact. Using complex forms makes it easy to compute variance at frequencies between harmonics. This is the subject of Section 1.5.6. One interesting result is that the variance at a nonharmonic frequency is uniquely related to the variances at all the harmonic frequencies. The seventh and last topic, in Section 1.5.7, is concerned with adding zeroes to a data set, why we might do that, and how to interpret the resulting spectrum

1.5.1 Aliasing, spectrum folding, and the Nyquist frequency

Aliasing is a direct consequence of digitally sampling an analog signal. Aliasing has no relevance to purely analog data records. To show how aliasing works, consider the three cases in Figure 1.26. In example (1) an analog sinusoidal wave with frequency 10 Hz is sampled at intervals of 0.1 s as indicated by the arrowheads. The dashed line connects the sample values. Based on just the sample values, we would likely (mistakenly) conclude the underlying signal has constant value. In example (2) there is a 9 Hz sinusoid sampled every 0.1 s. After fitting the sample values with a smooth line, we would likely (mistakenly) conclude that the underlying signal is a 1 Hz sinusoid. Example (3) indicates that for 0.1 s sampling, a 6 Hz sinusoid could just as well be interpreted as a 4 Hz sinusoid. These examples show that in digital sampling there is an inherent ambiguity in the frequency at which the true fluctuations are occurring. This is reflected in their line spectra shown in Figure 1.27. In example (1), all the variance in the true spectrum (solid bar) is at 10 Hz, but the observed spectrum indicates a nonvarying signal, that is, no variance at all. In example (2), the variance in the true spectrum is at 9 Hz while the observed spectrum (open bar) shows variance at 1 Hz. The true and observed spectra in

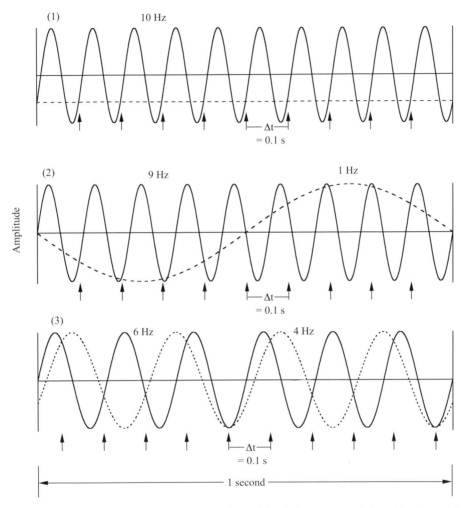

Figure 1.26 Three examples of aliasing indicated by dashed lines. (1) A 10 Hz sinusoid is sampled as a constant signal. (2) A 9 Hz sinusoid is sampled as a 1 Hz sinusoid. (3) A 6 Hz sinusoid is sampled as a 4 Hz sinusoid.

example (3) follow the same pattern as above. Another way to look at aliasing is that *more than two observations per cycle are required to unambiguously define a sinusoid.* Otherwise, it can be interpreted also as a sinusoid of lower frequency.

The picture that emerges from Figure 1.27 is that the calculated value of variance is folded about 5 Hz. This frequency is called the *folding* or *Nyquist frequency*, f_v, the latter named after Harry Nyquist who did pioneering work in signal analysis (Nyquist, 1928). In general, the Nyquist frequency is determined by the sampling interval Δt, that is, $f_v = 1/(2\Delta t)$; in the example just discussed, $f_v = 5$ Hz.

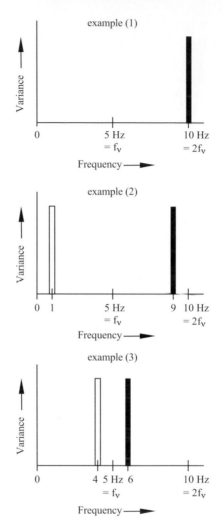

Figure 1.27 The true and observed spectra corresponding to the three examples in Figure 1.26. In each example above, the true spectrum is indicated by a solid bar and the observed spectrum by an open bar.

Furthermore, it is easy to conclude that the spectrum will repeat itself at frequency intervals of $\pm i/\Delta t$, $i = 1, 2, \ldots$. As an illustration, consider time series 1 given by

$$x_{1n} = \cos(2\pi f_m n\Delta t + \phi) \tag{1.39}$$

in which the series represents digital sampling of a sinusoid with frequency f_m, data point number n, and phase shift ϕ at intervals of Δt. Then define time series 2 as

$$x_{2n} = \cos[2\pi(f_m \pm i/\Delta t) n\Delta t + \phi]. \tag{1.40}$$

That is, time series 2 is digitally sampled in the same way as time series 1, except the frequency of the signal being sampled has been increased or decreased by integer multiples of twice the Nyquist frequency. Time series 2 can be expanded as a standard trigonometric angle-sum relation and then, because $\cos(2\pi in) = 1$ and $\sin(2\pi in) = 0$, reduced to

$$x_{2n} = \cos(2\pi f_m n \Delta t + \phi) \qquad (1.41)$$

the same formula as for time series 1. As a consequence of digital sampling, the x_{1n} and x_{2n} time series are identical, despite the fact that the underlying signals being sampled are different. Thus the same variance will be computed at f_m, $f_m \pm 1/\Delta t$, $f_m \pm 2/\Delta t$, and so on. We notice that negative frequencies are allowed. This is purely a mathematical convenience. While employing a spectrum that has both positive and negative frequencies is especially helpful in understanding aliasing, the "two-sided" spectrum concept also will be used later in Sections 1.5.5–1.5.7. In these sections we will find that mathematical formulas for spectra are more compact and easier to interpret when they include variance at both positive and negative frequencies.

Figure 1.28a summarizes aliasing from a schematic viewpoint. The *aliased spectrum* extends across *all* negative and positive frequencies with the spectrum repeated at intervals of $2f_v = 1/\Delta t$. In the jargon of spectrum analysis, the band of

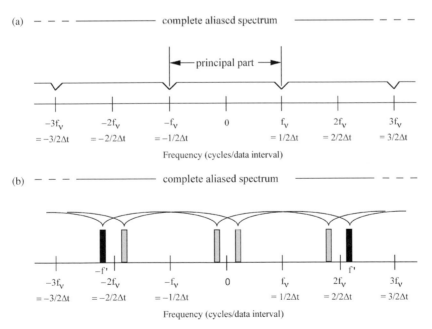

Figure 1.28 (a) The complete aliased spectrum and its principal part. f_v is the Nyquist frequency. (b) In a two-sided spectrum, one-half the variance appears at f' and one-half at $-f'$ and each is aliased to frequencies $\pm i/\Delta t$ from $\pm f'$ where i is an integer.

frequencies between $-f_\upsilon$ and $+f_\upsilon$ is called the *principal part* of the aliased spectrum. In practice, only the principal part is needed because the spectrum is repeated every $2f_\upsilon$ or $1/\Delta t$; that is, the principal part contains all the variance in the time series. Further insight into aliasing can be obtained by considering an input sinusoid with frequency greater than f_υ. Let us use the same frequency scale in Figure 1.28b and place the variance at f' between $2f_\upsilon$ and $3f_\upsilon$. Because we are using both positive and negative frequencies, the total variance at f' is split so that one-half the variance of the sinusoid is at f' and one-half at $-f'$. Figure 1.28b shows the solid vertical bars; their sum is the total variance. From Equations 1.40 and 1.41 the variances will be distributed to the open bars at frequencies $\pm i/\Delta t$ relative to $\pm f'$ as shown by the lines and pointers. If the input frequency happens to be a multiple of f_υ, no variance will appear at any frequency in accord with example (1) in Figure 1.27.

The repetition of the variance distribution in the principal part of the aliased spectrum in the remainder of the aliased spectrum is evident. If your preference is to deal only with variances at positive frequencies from 0 to f_υ, simply fold the spectrum from 0 to $-f_\upsilon$ around the origin from 0 to f_υ and add the variances.

It should be clear by now that it is important to know whether you are working with a two-sided $(-f_\upsilon$ to $f_\upsilon)$ spectrum or a one-sided spectrum (0 to f_υ) to get the correct total variance of the time series. In the former the total variance resides between $-f_\upsilon$ and f_υ while in the latter between 0 and f_υ. The variances at positive and negative frequencies in the former spectrum are one-half those in the latter. In the periodogram or Fourier analysis in previous sections, including Table 1.1, the harmonic variances were calculated at positive frequencies only, that is, from 0 to f_υ.

Let us return to example (2) in Figure 1.26 to create a new analog time series that is the sum of the original 9 Hz sinusoid and the 1 Hz aliased sinusoid (dashed line) and sample it at $\Delta t = 0.1$ s as shown. The value at each sample point will be twice as large (negative or positive) as in the original 9 Hz sinusoid. As a consequence, the observed variance at 1 Hz in the principal part of the aliased spectrum will be four times larger than that with only the original 9 Hz sinusoid. With a 180° phase shift of either wave (flip either sinusoid about the horizontal axis), the value at each Δt of the sum waveform is zero, and thus the variance is zero. In a word, this is why we have to be concerned with the effects of aliasing; variance at frequencies greater than f_υ will alter the true variance present at frequencies less than f_υ and produce an erroneous picture of variance. The seriousness of aliasing is in proportion to the ratio of the variance at frequencies outside the principal part to the true variance in the principal part.

Consideration of the potential for aliasing is critical to effective experiment design and proper analysis of results. To minimize aliasing, the sample rate should be such that practically all the variance will be at frequencies less than $1/(2\Delta t)$. If the general structure of the spectrum is unknown before sampling, experimentation may be required with different sampling rates to observe spectrum changes. If, for a given

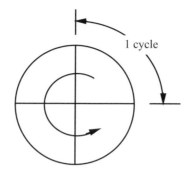

Figure 1.29   A four-spoke wagon wheel.

sampling rate, the potential exists for serious aliasing and the sampling rate cannot be increased, then one must filter the variance at frequencies $>1/(2\Delta t)$ before sampling. There is no effect on aliased variance if filtering is performed after digitizing. That is, the analog signal must be filtered.

A visual example of aliasing as seen in Western cowboy movies is the familiar changing of the direction of rotation of a wagon wheel as the wagon increases its speed from rest. The digital sampling is done by the camera shutter opening and closing 24 times each second.

Consider the four-spoke wheel in Figure 1.29. One cycle means rotation of the wheel $\frac{1}{4}$ revolution. When the wheel turns slowly, we see a continued forward rotation of the set of four spokes because there are many samples (shutter openings and closings) for the small angular rotation. As the wheel rate of rotation increases, the angular separation between successive samples also increases until the separation reaches $\alpha = 45°$, or $\frac{1}{2}$ cycle or $f_v = 0.5$ cycle/Δt, where $\Delta t = (1/24)$ s. This is the maximum observable frequency or rate of rotation of the wheel and is shown in Figure 1.30a. As the rate of rotation or actual frequency f increases beyond f_v, the observed frequency will be negative. This can be understood by referring to Figure 1.30b, keeping in mind that the sampling rate is fixed. Since $\alpha > 45°$, it is apparently easier for our brain to think the wheel has rotated not through angle α

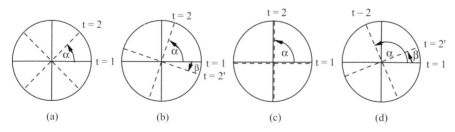

Figure 1.30   Successive positions of the four-spoke wagon wheel at times $t = 1$ and $t = 2$ for an increasingly higher rate of rotation from case (a) to case (d).

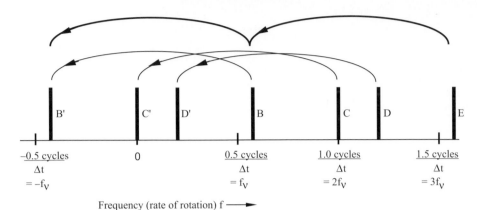

Figure 1.31   The aliased spectrum for frequency of rotation greater than the Nyquist frequency.

from position $t = 1$ to position $t = 2$, but through smaller angle β from position $t = 1$ to position $t = 2'$. (Of course, the appearance of the wheel is identical at $t = 2$ and $t = 2'$.) In the spectrum in Figure 1.31 this corresponds to the variance at frequency B aliased to frequency B'. As the wheel rotates faster, $\alpha = 90°$ and it appears that there is no motion (Figure 1.30c). Each spoke advances $1/4$ revolution or one cycle each time the shutter opens. In the spectrum this corresponds to variance at frequency C aliased to frequency C' (the origin). From B' to C' it appears that the wheel rotation rate is decreasing, that is, becoming less negative.

An increasing forward rate of rotation occurs for $\alpha > 90°$. For example, the variance at frequency D in Figure 1.31 is aliased to D'. As seen in Figure 1.30d, it is, again, apparently easier for our brain to accept rotation of the wheel through angle β at apparent time $t = 2'$ rather than larger angle α at real time $t = 2$. As the wheel rotates still faster, say at rate E in the spectrum, the rotation rate will be aliased back to frequency $E' = B'$, as shown by the heavy lines. Thus, as the rate of rotation of the wheel increases from zero it reaches a maximum forward rate, which instantaneously becomes the maximum backward rate, which then decreases to zero and the cycle starts all over again – all because of digitally sampling an analog signal.

1.5.2 Spectrum windows

Consider the "idealized time series" of 36 consecutive values of temperature at Phoenix, Arizona, during fair weather shown in Figure 1.32a. Because Phoenix is located in the southwestern desert of the United States, we expect a strong diurnal variation in air temperature. It is idealized because other harmonics that would normally contribute to the daily temperature variation, such as the semi-diurnal

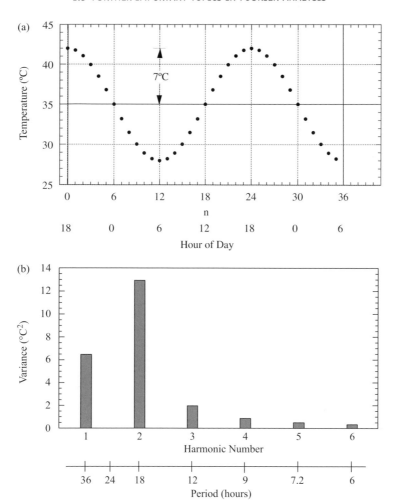

Figure 1.32 (a) Idealized time series of temperature at Phoenix, Arizona, during fair weather in July. (b) Periodogram to harmonic 6 of idealized time series of temperature in (a).

component seen in Figure 1.15, have been ignored. In a periodogram of this time series, the diurnal variation would occur at a nonharmonic frequency midway between harmonic 1 (fundamental period = 36 hours) and harmonic 2 (period = 18 hours). The purpose of this section is to show how the input variance at a nonharmonic frequency (the daily cycle here) gets distributed to the harmonic frequencies.

Apart from the constant offset value of 35 °C, the time series in Figure 1.32a is given by the sinusoid

$$x_n = a \cos(\omega n \Delta t - \phi), \quad n = 0, 1, \ldots, N-1$$

where x represents temperature, $\Delta t = 1$ hour, amplitude $a = 7\,°C$, data length $N = 36$ hours, phase angle $\phi = 0°$, and angular frequency $\omega = 2\pi \times 1.5/N$. That is, there are 1.5 cycles over the 36-hour record. Changing to angular frequency is merely a convenience to reduce the number of symbols in each equation. Figure 1.32b is the resulting periodogram in the form of a line spectrum out to harmonic 6. Harmonics 1 and 2, which are adjacent to the frequency of the input wave, account for about 80% of the variance of x_n; the higher harmonics account for the remaining variance. The big question is: How did the variance from the input wave get distributed to the various harmonics?

To find the answer, we first substitute x_n above into the equations for A_m and B_m (N even) in Table 1.1. Carrying out the summations is a tedious exercise in trigonometry, and the general procedure is shown in Appendix 1.D. We are really interested in the variance at harmonics, so the Fourier coefficients need to be squared according to $S_m^2 = (A_m^2 + B_m^2)/2$. This step is also given in Appendix 1.D, with the result that

$$
S_m^2(\omega) = \frac{a^2}{2} \left\{ \frac{\sin^2[N(\omega + \omega_m)/2]}{N^2 \sin^2[(\omega + \omega_m)/2]} + \frac{\sin^2[N(\omega - \omega_m)/2]}{N^2 \sin^2[(\omega - \omega_m)/2]} \right.
$$

$$
\left. + 2\cos[(N-1)\omega - 2\phi] \frac{\sin[N(\omega + \omega_m)/2]}{N\sin[(\omega + \omega_m)/2]} \times \frac{\sin[N(\omega - \omega_m)/2]}{N\sin[(\omega - \omega_m)/2]} \right\},
$$

$$
m \neq 0,\ N/2. \tag{1.42}
$$

It is not necessary to work through Appendix 1.D at this time. It is important, though, to be able to properly interpret Equation 1.42, and there are two ways. In the first way, $S_m^2(\omega)$ gives the variance at harmonic numbers m, where $0 < m < N/2$, due to an input sinusoid of amplitude a at angular frequency ω. Figure 1.32b is an example. In short, the equation shows how input variance $a^2/2$ is distributed among the harmonic frequencies.

The second way to interpret Equation 1.42 is to consider fixing m successively at $1, 2, 3, \ldots$, where $\omega_m = 2\pi m/N$, and, then, for each m, allow the input frequency ω to vary continuously over the range of frequencies in the spectrum. A plot of the ratio $S_m^2(\omega)/(a^2/2)$ for each m provides the "window" through which the spectrum is viewed at that harmonic for input variance at any ω. Figure 1.33 shows the spectrum window, that is, the part of Equation 1.42 in braces, for harmonics $m = 1, 2$, and 3 for a cosine input ($\phi = 0°$). The heavy solid line shows the location of the input wave at harmonic 1.5. To get the variance at harmonic 2, we multiply the variance of an integer number of cycles of the input wave, $a^2/2 = 24.5\,°C^2$, by 0.5277, the amplitude of the window associated with harmonic 2 (i.e., center curve) at the input frequency. The product is the variance at harmonic 2 in Figure 1.32b. The product of 0.2641,

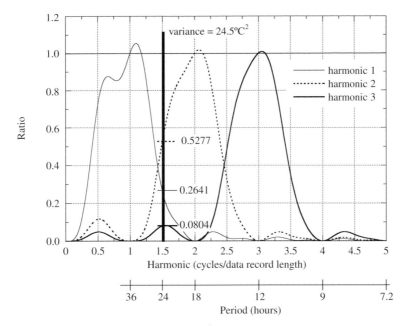

Figure 1.33 Spectrum windows from Equation 1.42 centered at harmonics 1, 2, and 3 for a cosine wave input. The product of the variance (24.5 °C²) in Figure 1.32a and the intersection of the spectrum windows yields the observed variance at harmonics 1, 2, and 3 in Figure 1.32b.

the amplitude of the window associated with harmonic 1 (i.e., left-hand curve) at the input frequency, and 24.5 °C² is the value of variance at harmonic 1 and, similarly, the variance at harmonic 3 is the product of 24.5 °C² and 0.0804 (right-hand curve). The windows for sine wave inputs ($\phi = 90°$) are shown in Figure 1.34; their individual shapes tend to be a reverse image of those in Figure 1.33. We conclude that the spectrum window depends on harmonic number and phase angle for a given N.

The spectrum window for the mean squared value standardized by the input variance is, from Equation 1.D.4,

$$A_0^2/(a^2/2) = 2\cos^2[(N-1)(\omega/2) - \phi]\frac{\sin^2(N\omega/2)}{N^2\sin^2(\omega/2)}. \tag{1.43}$$

Figure 1.35 shows the spectrum windows for a cosine input ($\phi = 0°$) and a sine input ($\phi = 90°$). For the Phoenix example ($\phi = 0°$) the value of the window at harmonic 1.5 is 0.001543, so that with a $= 7$ °C, $A_0 = 0.1944$ °C (to get the mean of the time series in Figure 1.32a, add back 35 °C). That A_0, the mean of the time series, is not zero is because the Phoenix time series does not have an integer number of cycles. In fact, if A_0 is multiplied by N $= 36$, the number of data, the

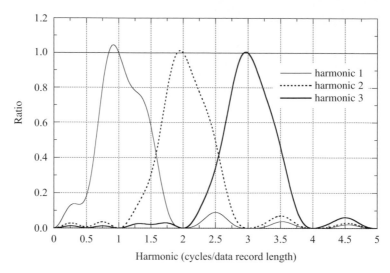

Figure 1.34 Spectrum windows (Equation 1.42) centered at harmonics 1, 2, and 3 for a sine wave input.

result is $7\,^\circ$C, the amplitude of the sinusoid. By matching positive departures from $35\,^\circ$C with negative departures, we see that only one of the two maximum positive values of temperature has a negative equivalent. At the Nyquist frequency, where $m = N/2$, $S^2_{N/2}(\omega)$ can be obtained directly from Equation 1.42 by dividing the right-hand side by two.

In general, the window shape is dependent on harmonic number, the number of data, and the phase angle of the input. When the number of data in a sample is

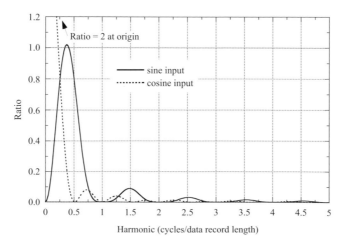

Figure 1.35 Spectrum windows (Equation 1.43) at the 0-th harmonic for sine and cosine inputs.

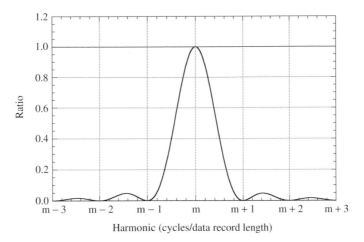

Figure 1.36   The spectrum window (Equation 1.44) at general harmonic m, when m is away from the low and high frequency ends of the periodogram.

$N \approx 100$ or higher, the main and adjacent lobes at the interior harmonics can be quite accurately modeled by simplifying Equation 1.42 to

$$S_m^2(\omega) = \frac{a^2}{2} \frac{\sin^2[N(\omega - \omega_m)/2]}{[N(\omega - \omega_m)/2]^2} \qquad (1.44)$$

which is dependent only on the number of data and the difference between the input frequency and the harmonic where the calculation is made. The maximum error in using Equation 1.44 in place of Equation 1.42 for interior harmonics is about $\pm 3\%$ for $N = 100$. This formula is the square of the familiar "diffraction function" common in optics and is plotted in Figure 1.36.

Assuming N is sufficiently large, we can think of the variance computed at a given harmonic frequency as the integral over the frequency range in the spectrum of a weight function (Equation 1.44) centered at that harmonic times an underlying, but unknown, spectrum. This process is repeated at all harmonics and results in variance "leaking" from one part of the spectrum to other parts of the spectrum. The variance observed at a particular harmonic does not necessarily mean that the data contain a pure tone at that harmonic. To find the variance of the Phoenix diurnal temperature cycle in a periodogram, a record length that is a multiple of 24 hours should be selected.

1.5.3 Detecting a periodic signal by averaging spectra

If we were to average together periodograms of equal length realizations from the same random process, harmonic by harmonic, we expect the averaged periodogram would be smoother than any individual periodogram. If a deterministic signal is present, its magnitude should not be affected by averaging. In this section we use the idea of averaging to investigate the particular problem of detecting a sinusoid embedded in white noise when multiple realizations are available.

The average of the two periodogram random variables $C_1(f_m)$ and $C_2(f_m)$ from realizations of equal length N (even) of a white noise process is

$$\overline{C}(f_m) = \frac{C_1(f_m) + C_2(f_m)}{2}$$

so that

$$\frac{\overline{C}(f_m)}{\Gamma(f_m)} = \frac{1}{4}\left(\chi_2^2 + \chi_2^2\right) = \frac{\chi_4^2}{4}$$

or, in general, averaging u spectra, $u = 1, 2, \ldots$, yields

$$\frac{\overline{C}_u(f_m)}{\Gamma(f_m)} = \frac{\chi_{2u}^2}{2u} \tag{1.45a}$$

for the interior harmonics, and

$$\frac{\overline{C}_u(f_m)}{\Gamma(f_m)} = \frac{\chi_u^2}{u} \tag{1.45b}$$

for the 0-th and Nyquist frequencies, that is, f_0 and $f_v = f_{N/2}$, respectively.

In parallel with Equation 1.30 we can use Equation 1.45a to determine the confidence interval for the population variance at the interior harmonics given the sample variance. Thus,

$$\Pr\left\{\frac{\overline{C}_u(f_m)}{\chi_{2u}^2(1 - \alpha/2)/2u} \leq \Gamma(f_m) \leq \frac{\overline{C}_u(f_m)}{\chi_{2u}^2(\alpha/2)/2u}\right\} = 1 - \alpha, \quad f_m \neq f_0, f_v$$

where α is the significance level. The interval between

$$2u\frac{\overline{C}_u(f_m)}{\chi_{2u}^2(1 - \alpha/2)} \quad \text{and} \quad 2u\frac{\overline{C}_u(f_m)}{\chi_{2u}^2(\alpha/2)}$$

is the $100(1 - \alpha)\%$ confidence interval for $\Gamma(f_m)$. By taking logarithms of the lower and upper limits of the confidence interval for $\log \Gamma(f_m)$, they become, respectively,

$$\log\overline{C}_u(f_m) + \log\left[\frac{2u}{\chi_{2u}^2(1 - \alpha/2)}\right] \quad \text{and} \quad \log\overline{C}_u(f_m) + \log\left[\frac{2u}{\chi_{2u}^2(\alpha/2)}\right].$$

As noted in Section 1.4.4, the width of the confidence interval will remain constant regardless of frequency. For example, consider averaging three Fourier spectra so that $u = 3$. The 95% ($\alpha = 0.05$) confidence interval for $\log \Gamma(f_m)$, $f_m \neq f_0$, f_v, extends from $\log \overline{C}(f_m) + \log (0.415)$ to $\log \overline{C}(f_m) + \log (4.85)$. In a similar manner, Equation 1.45b can be used to find the confidence interval for the population variance at the exterior harmonics.

If there are deterministic components in the spectrum, they will remain unchanged by spectrum averaging. Looking at this method in another way, averaging spectra can be used to detect deterministic components.

Consider the following computer simulation of

$$x_n = b \sin(2\pi f n - \phi) + \varepsilon_n, \quad n = 1, 2, \ldots, N$$

where $b = \sqrt{2}$, $N = 32$, $f = 6.25/N$, ϕ is phase angle ($0 \leq \phi < 360°$), and ε_n is white noise with population variance $\sigma^2 = 5$. If the signal-to-noise variance ratio (SNR) is defined to be

$$SNR = \frac{\dfrac{b^2}{2}}{\dfrac{\sigma^2}{(N-1)/2}} \tag{1.46}$$

the ratio of the variance of the sinusoid to the white noise variance at an interior harmonic frequency, its value is 3.1.

Each realization of length 32 comprises computer generated normal white noise, ε_n, added to the sinusoid with a different value of ϕ. If we make the null hypothesis that the variance spectrum comes from a white noise process and rewrite the two-sided equation for the confidence interval for the population variance in the form for a one-sided test only, namely,

$$\Pr\left\{0 \leq \overline{C}_u(f_m) \leq \frac{\chi^2_{2u}(1-\alpha)\Gamma(f_m)}{2u}\right\} = 1 - \alpha \tag{1.47}$$

we can use this formula to obtain the *a priori* upper confidence limit for the distribution of the observed harmonic variances. That we are dealing with only the upper confidence limit is because we are interested in the possible existence of a sinusoid, the indication of which is a peak in the spectrum. Figure 1.37 shows the spectra for six realizations of 32 data each for harmonics 3–9. The lower dashed line in each realization is the average variance for the interior harmonics and the arrow indicates the input frequency of the sinusoid. The upper dashed line shows the 95%

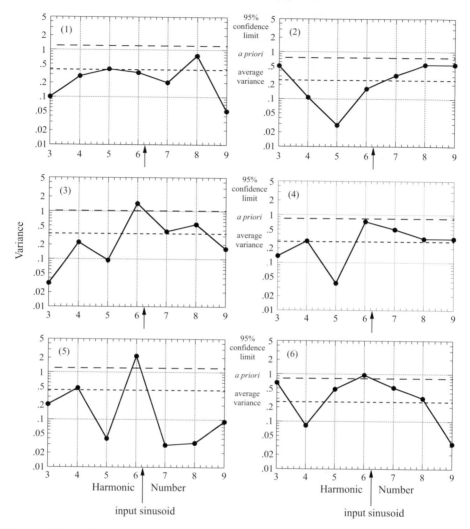

Figure 1.37 Periodograms of six realizations of a sinusoid plus white noise with signal-to-noise ratio (SNR) = 3.1 (solid line). The input sinusoid is 6.25 cycles over data length N = 32. The upper dashed line is the 95% *a priori* confidence limit. The lower dashed line is the average variance of the 15 interior harmonics for N = 32.

upper confidence limit for the observed variance of each individual realization (u = 1) and is computed from:

$$\frac{\chi_2^2(0.95)}{2}\Gamma(f_m)$$

where $\Gamma(f_m)$ is estimated by summing all of the harmonic variances (signal plus noise) of a realization and dividing by $(N-1)/2$. Realizations (1) through (6) of

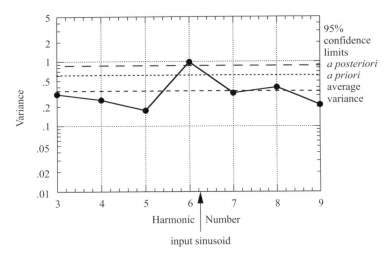

Figure 1.38 Average of the six periodgrams in Figure 1.37 (solid line). The upper two dashed lines are the *a priori* and *a posteriori* 95% confidence limits, the bottom dashed line is the average variance across all interior harmonics.

white noise yielded variances 6.72, 4.12, 5.60, 4.92, 6.77, and 4.25, respectively, compared to the population variance $\sigma^2 = 5$.

Figure 1.37 shows that the spectra vary considerably from one sample to the next, as expected, and that with a signal-to-noise ratio slightly greater than three, one may very well not detect the sinusoid using a single realization. Figure 1.38 shows the results of averaging the six spectra in Figure 1.37. The *a priori* 95% upper confidence limit is computed from $\chi^2_{12}(0.95)\Gamma(f_m)/12 = 0.61$, where the estimate of $\Gamma(f_m) = 0.348$ is obtained by averaging the estimates from all six realizations. The upper confidence limit occurs at a lower value of variance and closer to the mean than for any single periodogram. The result is the variance at harmonic 6 now clearly stands out.

Let us assume that the six realizations are actual data. Whether we would conclude that there is a significant oscillation at or near harmonic 6 depends on what we know from physical considerations may be occurring there and the likelihood the peak could have occurred by chance. With regard to the latter, we expect to observe, on average, 1 in 20 harmonic variances that exceed the *a priori* upper confidence limit. It is appropriate then to find the 95% *a posteriori* upper confidence limit, so that there is only a 5% chance that any one or more of the 15 harmonic variances will exceed this limit.

Following the procedure in Section 1.4.6, we divide $\alpha = 0.05$ by 15, the number of interior harmonics, the result being 0.0033. Next we find (estimate) from a chi-square table the abscissa of a χ^2_{12} distribution such that the area to the left is 0.9967. Thus, $\chi^2_{12}(0.9967)/12 = 2.46$. Given the above estimate of $\Gamma(f_m)$, the 95% *a posteriori* upper confidence limit is 0.86. Assume further that there is physical

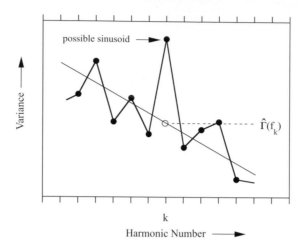

Figure 1.39 When the white noise null hypothesis is inadequate, it may be advantageous to fit a smooth curve or, as shown above, a straight line to the harmonic variances in the neighborhood of the possible sinusoid. A white noise test can be applied to the departures from the fitted curve.

evidence for a sinusoid close to harmonic 6. Since the white noise null hypothesis has been rejected, the variance in the sinusoid should have been removed from the estimate of $\Gamma(f_m)$, the consequence being the correct *a posteriori* upper confidence limit would be even lower. Therefore, the test is conservative in that if the null hypothesis is rejected at some value of α using the preceding method, the actual value of α is even less.

When the spectrum of random data is not white noise, the estimation of $\Gamma(f_m)$ in Equation 1.47 must be made at the frequency at which a sinusoid is suspected. One way to do this is to apply a straight-line fit to the surrounding periodogram values and use the value of the straight line at f_m as the estimate of $\Gamma(f_m)$, as illustrated in Figure 1.39. Confidence limits for the observed variance then can be computed if the departures from the straight line are suitably white. Another way is to model the underlying stochastic process using an appropriately smooth function and determine the white noise confidence limits with respect to the departures from the model. Crowley, Duchon, and Rhi (1986) show an example of the latter in which they searched for potential solar cycles in annual varve data.

We conclude this section by saying that if one were fortunate enough to have six or more realizations from a random process in which there is a deterministic sinusoid with the signal-to-noise ratio of three or greater, there is a good chance of detecting the sinusoid in the averaged periodogram. The determination of the minimum SNR required for detection and statistical confirmation of a sinusoid, in general, is complicated because the outcome depends on its proximity to the nearest harmonic and the spectrum of the noise. For example, if the sinusoid is located between harmonics, the spectrum window will distribute its variance to a number of harmonics (Section 1.5.2). If the noise spectrum falls or rises sharply where it is located, we expect that the SNR would have to be very large in order to detect a pure sinusoid.

1.5.4 Effect of data length on detecting a periodic component

As noted in Section 1.4.5, the Fourier spectrum of random data can be viewed as an "unstable" spectrum because increasing the data length does not reduce the variability of variance computed at any harmonic from one realization to the next. Rather, an increase in data length results in an increase in frequency resolution; if the data length is doubled, the bandwidth or frequency separation between adjacent harmonic frequencies is halved. The dof for each approximately independent variance estimate is still two; as a result, the statistical distribution of each variance ratio $C(f_m)/\Gamma(f_m)$ remains $\chi_2^2/2$.

As in the previous section, consider a sinusoidal signal to which is added white noise. In this case let the signal have an integer number m cycles. The variance at the harmonic of the sinusoid is the sum of three terms: the variance of the sinusoid, the variance of the random component, and the covariance between the two harmonics – one from the sinusoid, the other from the noise. This can be understood by considering the sample variance of the sum of two sinusoids $x_{1n} = a_1 \cos(2\pi mn/N + \varphi_1)$ and $x_{2n} = a_2 \cos(2\pi mn/N + \varphi_2)$, where x_{1n} is the sinusoid and x_{2n} the Fourier component of random noise at the frequency of the sinusoid. It can be shown that the variance of the sum

$$S^2(x_{1n} + x_{2n}) = (1/N) \sum_{n=1}^{N} (x_{1n} + x_{2n})^2$$

reduces to

$$S^2(x_{1n} + x_{2n}) = (a_1^2 + a_2^2)/2 + a_1 a_2 \cos(\varphi_1 - \varphi_2). \tag{1.48}$$

A convenient way to prove Equation 1.48 is to express x_{1n} and x_{2n} in terms of complex exponentials using Euler's formula and then apply the summation procedure in Appendix 1.B – similar to the way it was applied it in Section 1.2.2. We see from Equation 1.48 that, depending on the magnitudes of a_2 and φ_2 in a particular realization of noise, the variance at the harmonic of the sinusoidal signal could be larger or smaller than the variance of the sinusoid itself. In an expected sense there is no preference for the covariance term in Equation 1.48 to be either positive or negative because a_2 has no sign preference and is uncorrelated with φ_2.

The important point to remember is that if the data length is doubled, the white noise variance will be distributed over twice as many frequencies and, on average, reduced by a factor of two at the frequency of the sinusoidal signal. The variance of the sinusoid will remain unchanged but occurs at twice the original harmonic number. Of course, there will be a proportional reduction in white noise at any harmonic for other integer multiple increases in data length.

If the sinusoid is not at a harmonic frequency, the most likely case in practice, the results are more complex, but in the expected sense can be qualitatively inferred from multiplying the spectrum window with the sinusoidal input variance, as discussed in Section 1.5.2. For example, if the frequency of the sinusoid lies midway between adjacent harmonics in the periodogram, the variance at the same frequency after doubling the data length will contain all the variance of the sinusoid. Considering only the variance of the sinusoid, its value will be more than twice the values at the adjacent harmonics in the original periodogram because the spectrum window spreads the variance of the sinusoid to *all* harmonics, not just the two adjacent harmonics. If the sinusoid is nearer to one or the other adjacent harmonics in the original periodogram, the variance will be mostly contained in the two harmonic frequencies that surround it in the spectrum for the case of twice the original data length. In summary, increasing the length of a time series that is stationary and contains a deterministic component results in improved ability to distinguish the variance of the deterministic component from the surrounding harmonic variances. At an appropriate stage, one can test for statistical significance of a possible sinusoid using one of the approaches given in Section 1.4.6.

As an example, consider the simulated time series $b \sin(2\pi ft - \pi/4)$, where $b = \sqrt{2}$ and $t = 1, 2, \ldots, N$, to which white noise is added. The white noise has a population variance $\sigma^2 = 5.0$. In Figure 1.40, curve (a) shows the distribution of percentage of total variance for harmonic numbers three through nine for the sinusoid plus a realization of white noise for $N = 32$ and $f = 6/N$. The signal-to-noise ratio, as defined by Equation 1.46, is 3.1. Percentage of total variance is used as the ordinate so that comparisons between realizations are not affected by varying amounts of total sample variance. Curve (a) indicates that the periodogram estimates vary considerably and that there would be no reason to expect an input sinusoid at harmonic 6. The periodogram of the white noise by itself (not presented) shows that, by chance, the value of variance at harmonic 6 is small compared to the adjacent variances and the phase angle between the component of white noise at harmonic 6 and the sinusoidal signal is about 30°. The combination of these two factors yields the value shown at harmonic 6 as dictated by Equation 1.48.

Curve (b) in Figure 1.40 results from extending the sinusoid and the realization of white noise associated with curve (a) to double their lengths. That is, the seed for the white noise random number generator was the same for curves (a) and (b). As expected, the periodogram values are generally reduced in magnitude as the

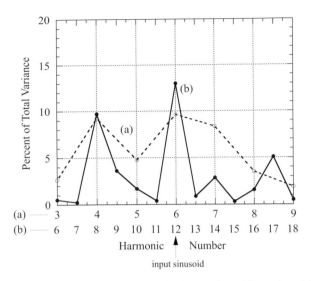

Figure 1.40 Partial periodograms for sinusoidal input plus white noise. (a) For signal-to-noise ratio (SNR) $= 3.1$ and data length $N = 32$. (b) For SNR $= 6.2$ and $N = 64$.

white noise variance is now distributed across twice the number of harmonics in (a). The presence of an input sinusoid is more in evidence with a SNR $= 6.2$, double that in (a) (ignoring the -1 contribution in the denominator of Equation 1.46). Even with this higher SNR, there is still considerable variance at harmonic 8.

Figure 1.41 is similar to Figure 1.40 with two exceptions: the input sinusoid is at harmonic 6.25 in the 32-point data set and there is a second doubling of the initial data length to yield a 128-point data set. The highest peak in curve (a) occurs at harmonic 7 with a comparatively small value at harmonic 6. By chance, the component of white noise at harmonic 6 is nearly out of phase ($\sim 170°$) with the component of the input sinusoid at harmonic 6 (negative covariance term). At the same time, there is only about a $70°$ phase difference between the component of white noise at harmonic 7 and the component of the input sinusoid at harmonic 7 (positive covariance term). The result is that the contribution of the input sinusoid to the variance at harmonic 7 is about 1.5 times greater than that at harmonic 6. This, coupled with the much larger variance of the random component at harmonic 7 than at harmonic 6, yields the magnitudes shown. Curve (b) shows the spectrum when the data are extended to twice the original length so that the SNR is 6.2. Generally, the periodogram values are less than those in (a), as anticipated. The frequency of the input sinusoid now lies midway between adjacent harmonics. The phase differences between the components of white noise and the input sinusoid at harmonics 12 and 13 and the magnitudes of the components account for the similar percentages of total variance at these harmonics. Other large percentages of total variance occur at harmonics 8, 14, and 17, due mainly to the strength of the

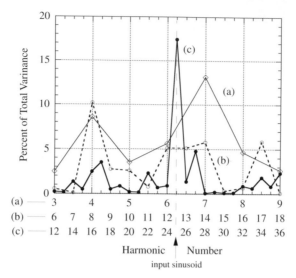

Figure 1.41 Partial periodograms for sinusoidal input plus white noise. (a) For signal-to-noise ratio (SNR) = 3.1 and data length N = 32. (b) For SNR = 6.2 and N = 64. (c) For SNR = 12.4 and N = 128.

component of white noise as leakage of variance from the input sinusoid diminishes with distance from harmonic 12.5.

In curve (c) in Figure 1.41, the data have been extended to four times the original length, the SNR thus being 12.4. From Table 1.7, which applies to curve (c), the phase difference between the input sinusoid and the sinusoid of noise at harmonic 25 is about 95°. In the "worst case" that could have arisen, the phase angle difference would be 180°, with the result that the ordinate would have been 9.4%. This figure is not too different from the values of 7.3% and 8.6% that were found at harmonics 2 and 10 (not shown), in which situation there would be little evidence for the deterministic component at harmonic 25. The figure of 9.4% can be calculated from values in the total variance and variance columns of Table 1.7 and Equation 1.48. The calculation is a good exercise to demonstrate understanding of how periodogram variances can change due to phase angle differences when two sinusoids are summed.

Now we can adapt Equation 1.47 to the ordinate in Figure 1.41 in order to find the $100(1 - \alpha)\%$ *a priori* confidence interval. Dividing each term by the total variance, such that the confidence interval is expressed as a percentage, yields

$$\Pr\left\{0 \le \frac{C(f_m)}{5.882} \le \frac{\chi_2^2(1 - \alpha)\Gamma(f_m)}{2 \times 5.882}\right\} = 1 - \alpha. \qquad (1.49)$$

The white noise estimate for $\Gamma(f_m)$ is $5.882/((N - 1)/2) = 2 \times 5.882/127 = 0.0926$. The resulting upper limit of the 95% *a priori* confidence interval is 4.7%. Among the

Table 1.7   Statistical properties of signal $\sqrt{2} \sin(2\pi\, 25\, n/128 - \pi/4)$ and a realization of a time series of 128 values of white noise.

Time series	Total variance	Variance at harmonic 25	Percentage of total variance	Phase angle (degrees)
signal	1.000	1.000	100.00	135.0
noise	4.936	0.084	1.70	39.6
signal + noise	5.882	1.030	17.51	118.5

63 interior harmonics there are four with values outside this limit. These include harmonic 25 for both the observed (95°) and "worst case" (180°) phase differences. On average, one would expect about three (63×0.05) rejects under the white noise null hypothesis. Using only the *a priori* confidence limit, it is unclear whether to reject or not reject the white noise null hypothesis that the data set comes from a white noise process. Accordingly, we calculate the 95% *a posteriori* confidence interval from Equation 1.49 after replacing the argument of χ_2^2 by $1 - \alpha/63 = 0.99921$. Integrating Equation 1.36 between 0 and the upper limit of integration results in $\chi_2^2(0.99921) = 14.278$. The upper limit of the *a posteriori* confidence interval is, therefore, 11.2%. The only harmonic whose percentage of total variance exceeds this limit is that at harmonic 25 for the 95° phase angle difference case. There are no harmonics with values exceeding this limit for the 180° "worst case."

We conclude that even with a SNR as large as 12, it can be difficult, in general, to not only detect a sinusoidal signal in the presence of white noise in a given realization, but to show also that it is statistically significant. Factors that contribute to this difficulty are (i) the occurrence of the sinusoid between harmonics and the attendant spectrum window effects and (ii) the chance occurrence of a combination of amplitude and phase angle of random noise at the same harmonic as the signal that significantly cancels the signal variance.

To briefly summarize Sections 1.5.3 and 1.5.4, we can state that the detection of a sinusoidal signal embedded in noise will be enhanced by either increasing the data length (with resultant increase in the SNR) or averaging a number of periodograms (with resultant narrowing of the spectrum confidence interval). Increasing the data length forces the periodic component to be closer to a harmonic frequency.

1.5.5 Complex representation of Fourier series

The most compact expression of a Fourier synthesis is that written in complex exponential form. The purpose of this section is to develop complex exponential forms for Fourier synthesis and analysis producing what is called a *Fourier transform pair*. The periodogram is then expressed in terms of complex coefficients. We shall

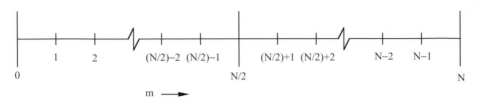

Figure 1.42   Fourier coefficents A_m and B_m computed at harmonics $(N/2) + 1$ to $N - 1$ can be exactly matched to Fourier coefficients computed at harmonics 1 to $(N/2) - 1$ with appropriate change in sign. N is even.

see that one of the features of the periodogram in complex form is that, with a couple exceptions, the Fourier coefficients are one-half the magnitude given in Table 1.1. The exceptions occur at the zero harmonic (N even or odd) and the Nyquist frequency (N even), in which cases there is no change of magnitude. The compactness also can result in an amplitude spectrum that includes both positive and negative harmonics (or frequencies), as described in Section 1.5.1 and Figure 1.28.

Figure 1.42 shows the locations of the harmonic coefficients A_m and B_m along a harmonic axis that has been extended to twice its usual length. If the range of m in the synthesis formula

$$x_n = A_0 + \sum_{m=1}^{\frac{N}{2}-1} \left(A_m \cos\frac{2\pi mn}{N} + B_m \sin\frac{2\pi mn}{N} \right) + A_{N/2} \cos\pi n \qquad (1.50)$$

from Table 1.1 for N even were to be extended beyond N/2, what would happen to the values of A_m, B_m, $\cos\frac{2\pi mn}{N}$, and $\sin\frac{2\pi mn}{N}$? Using trigonometric identities for the sum and difference of two angles, the results for the sine terms will be

$$\sin\frac{2\pi\left(\frac{N}{2} + m\right)n}{N} = \sin(\pi n)\cos(2\pi mn/N) + \cos(\pi n)\sin(2\pi mn/N) \qquad (1.51)$$

and

$$\sin\frac{2\pi\left(\frac{N}{2} - m\right)n}{N} = \sin(\pi n)\cos(2\pi mn/N) - \cos(\pi n)\sin(2\pi mn/N). \qquad (1.52)$$

Because the first term on the right-hand side in both equations is always zero and the second term is the same except for sign, it follows that

$$\sin\frac{2\pi\left(\frac{N}{2} + m\right)n}{N} = -\sin\frac{2\pi\left(\frac{N}{2} - m\right)n}{N}. \qquad (1.53)$$

In a similar manner it can be shown that

$$\cos\frac{2\pi\left(\frac{N}{2}+m\right)n}{N} = \cos\frac{2\pi\left(\frac{N}{2}-m\right)n}{N}. \tag{1.54}$$

Thus, the sine terms show odd symmetry about harmonic N/2 and the cosine terms show even symmetry. For N odd, the location of the Nyquist frequency is between adjacent harmonics that surround the Nyquist frequency. Nevertheless, the same pattern of symmetry about the Nyquist frequency holds for N odd as for N even.

Continuing with N even, since A_m and B_m both involve the cosine and sine terms above,

$$A_{\frac{N}{2}+m} = A_{\frac{N}{2}-m} \quad \text{and} \quad B_{\frac{N}{2}+m} = -B_{\frac{N}{2}-m}. \tag{1.55}$$

Thus, noting the even and odd symmetry of the Fourier cosine and sine coefficients, respectively, x_n can be written

$$x_n = \sum_{m=0}^{N-1}\left(A'_m \cos\frac{2\pi mn}{N} + B'_m \sin\frac{2\pi mn}{N}\right) \tag{1.56}$$

where $A'_m = A_m/2$ and $B'_m = B_m/2$, except $A'_0 = A_0$ (N even or odd) and $A'_{N/2} = A_{N/2}$ (N even). Fourier coefficients A_m and B_m are the original coefficients defined in Table 1.1. From this point forward, whenever a primed Fourier coefficient is observed, it means that its value is one-half the value of an unprimed coefficient, except as noted. Primed coefficients have been used already in Appendix 1.D.

Before rewriting the expression above in terms of complex numbers, let us briefly review what we mean by a complex number. A complex number is given by

$$z = x + iy$$

and its complex conjugate by

$$\overset{*}{z} = x - iy$$

where x and y are real numbers and i the *imaginary unit* defined by

$$i = \sqrt{-1}.$$

The real number x is called the *real part* of z (or its conjugate) and the real number y is called the *imaginary part* of z (or its conjugate). (Note that the x notation here is distinct from notation x_n for the time series above.) The complex number z can be

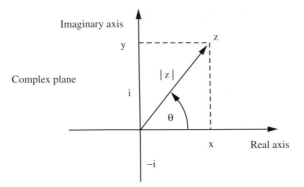

Figure 1.43   The complex plane.

easily interpreted as a vector in the complex plane shown in Figure 1.43 extending
from the origin to the intersection of x and y.
   The length of vector z is given by

$$|z| = \sqrt{x^2 + y^2}$$

and its direction by

$$\theta = \tan^{-1}\left(\frac{y}{x}\right).$$

Since

$$x = |z| \cos\theta \quad \text{and} \quad y = |z| \sin\theta$$

it is apparent that $z = x + iy$ can be written in the equivalent form

$$z = |z| \cos\theta + i|z| \sin\theta = |z| (\cos\theta + i \sin\theta)$$

which, from Euler's formula, can be written

$$z = |z| e^{i\theta}.$$

This is called the *polar* or *trigonometric* form of a complex number.
   As for now, we represent the Fourier coefficients using complex numbers in order
to rewrite the synthesis formula in complex exponential form. That is,

$$X_n = \sum_{m=0}^{N-1} (A'_m - iB'_m)\left(\cos\frac{2\pi mn}{N} + i\sin\frac{2\pi mn}{N}\right). \tag{1.57}$$

The cross product terms will vanish in the summation because of their odd symmetry
about $m = N/2$. For example, the product of A'_m with $\sin(2\pi mn/N)$ for $0 < m < N/2$

will be identical to the same product at harmonic $N - m$, but of opposite sign since A'_m is an even function about $N/2$ and $\sin(2\pi mn/N)$ is an odd function.

If we let $S'_m = A'_m - iB'_m$, then, from Table 1.1 and the Fourier coefficients there divided by two, as required earlier,

$$S'_m = \frac{1}{N}\sum_{n=0}^{N-1} x_n \cos\frac{2\pi mn}{N} - i\frac{1}{N}\sum_{n=0}^{N-1} x_n \sin\frac{2\pi mn}{N}$$

$$= \frac{1}{N}\sum_{n=0}^{N-1} x_n \left(\cos\frac{2\pi mn}{N} - i\sin\frac{2\pi mn}{N}\right), \quad 0 \le m \le (N-1),\ N\ \text{even.}$$

(1.58)

S'_m is the complex Fourier coefficient at the m-th harmonic frequency.

Using Euler's formula, expressions for x_n and S'_m can be expressed very compactly and symmetrically as

$$S'_m = \frac{1}{N}\sum_{n=0}^{N-1} x_n \exp(-i2\pi mn/N), \quad m = 0, 1, \ldots, N-1 \tag{1.59}$$

and

$$x_n = \sum_{m=0}^{N-1} S'_m \exp(i2\pi mn/N), \quad n = 0, 1, \ldots, N-1. \tag{1.60}$$

These equations constitute a *digital Fourier transform pair*, are valid whether N is even or odd, and could be written also

$$S'_m = \frac{1}{N}\sum_{n=0}^{N-1} x_n \exp(-i2\pi mn/N), \quad m = -[(N-1)/2], \ldots, 0, \ldots, [N/2]$$

$$\text{or } m = -[N/2], \ldots, 0, \ldots, [(N-1)/2]$$

(1.61)

and

$$x_n = \sum_{m=-[(N-1)/2]}^{[N/2]} S'_m \exp(i2\pi mn/N), \quad n = 0, 1, \ldots, N-1$$

or

$$x_n = \sum_{m=-[N/2]}^{[(N-1)/2]} S'_m \exp(i2\pi mn/N), \quad n = 0, 1, \ldots, N-1 \tag{1.62}$$

where [q] means truncation of q. The new limits on m follow from the easily proved relation $S'_{m \pm kN} = S'_m$, where k is an integer. Equations 1.60 and 1.62 are referred to as *inverse* Fourier transforms of Equations 1.59 and 1.61, respectively. Whenever there is a Fourier transform pair, the equation for the time or space function in terms of the frequency function is considered the inverse Fourier transform.

If the variance in the periodogram is denoted by C'_m, then

$$C'_m = S'_m \times S'^*_m = (A'_m - iB'_m)(A'_m + iB'_m) = A'^2_m + B'^2_m,$$

$$m = -[(N-1)/2], \ldots, 0, \ldots, [N/2] \qquad (1.63)$$

where the asterisk again indicates complex conjugate. Ordinarily, variance is not computed at $m = 0$. Notice the periodogram here is two-sided; that is, there are variances at both negative and positive harmonics. Their use is a mathematical convenience. Recall that the primed Fourier coefficients are one-half the values given in Table 1.1 except at $m = 0$ and $m = N/2$ (N even). To match the one-sided spectrum in Table 1.1 for N even or odd, the variances have to be doubled according to

$$S^2_m = C_m = C'_m + C'_{-m} = 2C'_m, \quad m \neq 0; \ m \neq N/2 \ (N \text{ even})$$

and   (1.64)

$$S^2_{N/2} = C_{N/2} = C'_{N/2}, \quad (N \text{ even}).$$

1.5.6 The spectrum at nonharmonic frequencies

It was pointed out in Section 1.2 that the total variance in a data set can be shown to be the sum of the variances at the harmonic frequencies. In terms of accounting for total variance, there is no need to examine the spectrum at a frequency resolution higher than the spacing between adjacent harmonics. Nevertheless, a value of variance can be computed at any frequency by changing the cosine and sine arguments in the spectrum formulations from $2\pi mn\Delta t/(N\Delta t)$ to $2\pi fn\Delta t$ where f is frequency (in cycles per time interval between samples). In this section we derive a formula from which we conclude that the variance spectrum $C'(f)$, available at a continuum of frequencies, is uniquely related to the variances at the harmonic frequencies C'_m.

Define $S'(f)$ to be the complex amplitude coefficient at frequency f. Then, by analogy with Equation 1.61,

$$S'(f) = \frac{1}{N} \sum_{n=0}^{N-1} x_n \exp(-i2\pi fn\Delta t), \quad -1/(2\Delta t) \leq f \leq 1/(2\Delta t). \qquad (1.65)$$

Now substitute the first form of Equation 1.62 for x_n to get

$$S'(f) = \sum_{m=-[(N-1)/2]}^{[N/2]} S'_m \sum_{n=0}^{N-1} \left(\frac{1}{N}\right) \exp\left[i2\pi\left(\frac{m}{N} - f\Delta t\right)n\right] \qquad (1.66)$$

where [q], as earlier, means the truncated value of q in the summation limits.
Using Equation 1.B.4 to obtain the second summation yields

$$S'(f) = \sum_{m=-[(N-1)/2]}^{[N/2]} S'_m \times E_m(f), \qquad -1/(2\Delta t) \le f \le 1/(2\Delta t) \qquad (1.67)$$

where

$$E_m(f) = \exp\left[i(N-1)\left(\frac{m}{N} - f\Delta t\right)\pi\right] \times \frac{\sin\left[\left(\frac{m}{N} - f\Delta t\right)N\pi\right]}{N\sin\left[\left(\frac{m}{N} - f\Delta t\right)\pi\right]}. \qquad (1.68)$$

Equation 1.67 tells us that each complex Fourier coefficient at a frequency between two adjacent harmonics is a weighted sum of the S'_m harmonic coefficients. This means that calculating Fourier coefficients at nonharmonic frequencies yields no additional insight into the variance structure of the data; all of the variance information is revealed by the harmonic coefficients.

If S'_m and $E_m(f)$ are replaced by their conjugates, the conjugate companion of Equation 1.67 will result and the weighted sum relation will apply to $\overset{*}{S'}(f)$. Because the periodogram is the product of S'_m and $\overset{*}{S'_m}$, we conclude that at any frequency, f, the variance spectrum

$$C'(f) = S'(f) \times \overset{*}{S'}(f), \qquad -1/(2\Delta t) \le f \le 1/(2\Delta t) \qquad (1.69)$$

is a weighted sum of the variances at the harmonic frequencies. When f is at a harmonic frequency, the weight function is zero at all other harmonics except the one under consideration.

Let us re-examine Equation 1.68 for the case when $m = 0$. Then,

$$E_0(f) = \exp[-i(N-1)\pi f\Delta t] \frac{\sin(\pi N f\Delta t)}{N\sin(\pi f\Delta t)}. \qquad (1.70)$$

Note that the limit of $E_0(f)$ is one as f tends to zero. If S'_0, the mean of the series, is large, the product of $E_0(f)$ and S'_0 will provide a large contribution to $S'(f)$ when f is in the neighborhood of the origin. The variance $C'(f)$ will then include a large contribution from the mean. Since it is the second moment about the mean that is desired, this

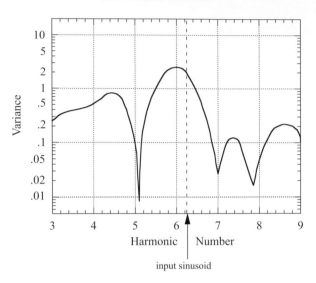

Figure 1.44 Spectrum of sinusoid with frequency 6.25 cycles over data length $N = 32$ with added white noise. The ratio of the signal variance to the white noise variance at an internal harmonic (SNR) is 3.1 (see Equation 1.46).

contribution must be deleted. Therefore, in applying Equation 1.69 it is to be understood that the sample mean has been removed before computing $S'(f)$ or $C'(f)$.

One might expect that $C'(f)$ could be used to find the exact frequency of a deterministic signal embedded in noise. Unfortunately, this is not true, and an illustration of this fact is shown in Figure 1.44. The spectrum here is the same as spectrum (5) in Figure 1.37, except that the variances were calculated at frequency increments corresponding to 1/20 the harmonic spacing using Appendix 1.A. Due mainly to the leakage of noise variance from surrounding harmonics, the peak in the spectrum occurs not at the frequency of the input sinusoid (harmonic 6.25) but slightly to the left of harmonic 6.

If the noise is reduced to zero, however, the input frequency can be accurately determined, as shown in Figure 1.45. Why is this? In the first place there is no noise to contend with and, therefore, no noise leakage. In the second place, for interior frequencies and sufficiently large N, the frequency at which the peak in the spectral window in Equation 1.42 occurs when $2\pi f_i$ is substituted for ω_m is close to the frequency of the input sinusoid f_i. Recall from Figures 1.33 and 1.34 that the peak in the spectrum window at harmonic 1 is displaced from harmonic 1 for a pure cosine or sine input. It was found from simulations that when $N > 100$ a reasonably accurate estimate of the frequency of an input sinusoid away from the ends of the spectrum can be made because the window is nearly symmetric and its peak is close to the center of the window. This parallels the earlier finding in Section 1.5.2 that the squared diffraction function in Equation 1.44 provides a good approximation to the window function in Equation 1.42 away from the spectrum ends for $N > 100$. Of course, the larger the value of N, the greater the accuracy. From Figures 1.44

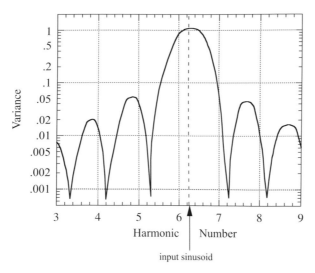

Figure 1.45 Spectrum of sinusoid with frequency 6.25 cycles over data data length N = 32.

and 1.45 we conclude that the higher the ratio of the deterministic signal variance to the noise variance, the more accurately one can estimate the frequency of the signal.

The various equations developed in this and the previous sections typically are not used in computing a periodogram. Instead, we use a straightforward algorithm employing the formulas in Table 1.1 or a fast algorithm as in the computer program in Appendix 1.A. The latter algorithm permits us to evaluate the Fourier coefficients and variances at as high a resolution in frequency as we wish. When we do this, we now know that the variance computed at an off-harmonic frequency is a weighted sum of all harmonic variances, and the closer they are to a given off-harmonic frequency, the greater their influence.

1.5.7 Padding data with zeroes

In this section we investigate a topic of practical interest wherein a time series with zero mean is modified by appending zeroes to it in order to obtain a desired length. The procedure is called "padding data with zeroes" and a common purpose is to match the length of a record with that required when using an FFT (fast Fourier transform) algorithm to analyze many and/or long data sets. As an example, if we had an 83 point sequence and were using a simple FFT requiring 2^k points, we could add 45 zeroes to obtain $2^7 = 128$. We will show that the periodogram of the padded data is identical to the variance spectrum (Equation 1.63) of the original data computed at intervals in frequency of $1/128$, except for a multiplicative constant. It may seem odd that one would add a sequence of zeroes to a time series (after removal of the mean) and compute a periodogram that has any meaning. The interesting aspect is that in calculating the Fourier coefficients, the padded series can be partitioned into the original series and the sequence of zeroes, the latter contributing nothing to the coefficients. The result is a spectrum with higher resolution than the period-ogram of the original data.

Let us begin by considering a time series of data and subtract its mean from each datum to get data set A. Next consider data set B. It is the same as data set A except that zeroes have been appended in order to apply an FFT. The means of data sets A and B are zero. The variances of both are the same if, for set B, the coefficient of the sum in the expression for variance is the same as that for set A – a condition that we now investigate.

The formula for the Fourier coefficients in data set A is, following Equations 1.58 and 1.59,

$$S'_m = A'_m - iB'_m = \frac{1}{N} \sum_{n=0}^{N-1} x_n \exp(-i2\pi mn/N), \quad m = -[(N-1)/2], \ldots, 0, \ldots, [N/2]$$

$$(1.71)$$

where m is harmonic number, x_n is the n-th datum, N is the number of data and [q] indicates truncated value, as before.

Of course, we know from the previous section that we can calculate Fourier coefficients at higher resolution in frequency than the harmonic frequencies (Equation 1.65) and they are completely dependent on those at the harmonic frequencies (Equation 1.67). Consider increasing the number of complex coeffi-cients N in set A by a factor R, such that RN is the number of data needed by an FFT. Then the new formula for the high resolution coefficients in data set A is

$$S'_r = A'_r - iB'_r = \frac{1}{N} \sum_{n=0}^{N-1} x_n \exp(-i2\pi rn/(RN)),$$

$$r = -[(RN-1)/2], \ldots, 0, \ldots, [RN/2]. \qquad (1.72)$$

To distinguish padded data set B from data set A, we will use bold notation, for example, $\mathbf{x}_n$. Data set B has RN data and, by analogy with Equation 1.71, noting that $\mathbf{x}_n = 0$ beyond $n = N - 1$, its Fourier coefficients are

$$S'_r = \mathbf{A}'_r - i\mathbf{B}'_r = \frac{1}{RN} \sum_{n=0}^{RN-1} \mathbf{x}_n \exp[-i2\pi rn/RN]$$

$$= \frac{1}{RN} \sum_{n=0}^{N-1} \mathbf{x}_n \exp(-i2\pi rn/RN) + \frac{1}{RN} \sum_{n=N}^{RN-1} \mathbf{x}_n \exp(-i2\pi rn/RN)$$

$$= \frac{1}{RN} \sum_{n=0}^{N-1} \mathbf{x}_n \exp(-i2\pi rn/RN), \quad r = -[(RN-1)/2], \ldots, 0, \ldots, [RN/2].$$

$$(1.73)$$

Except for the coefficient of the summation, Equation 1.73 is the same as Equation 1.72. If it were desired that the Fourier coefficients of padded data set B be the same as the high resolution coefficients of data set A, the former coefficients need to be multiplied by R.

Whether one uses padding (Equation 1.73) or direct calculation (Equation 1.72), the total variance derived from the Fourier coefficients must equal the variance in the data. The variance at a Fourier harmonic frequency can be obtained by forming the product $S_m \times S_m^*$, in which the asterisk means complex conjugate. Accordingly, from Equations 1.61 and 1.63,

$$
S_m' \times S_m'^* = A_m'^2 + B_m'^2 = \frac{1}{N^2} \left| \sum_{n=0}^{N-1} x_n \exp(-i2\pi mn/N) \right|^2,
$$

$$
m = -[(N-1)/2], \ldots, 0, \ldots, [N/2] \tag{1.74}
$$

the total variance of which is

$$
\sum_{m=-[(N-1)/2]}^{[N/2]} \left(A_m'^2 + B_m'^2 \right). \tag{1.75}
$$

When variances are computed at a higher resolution in frequency than that associated with just the harmonic frequencies, they must be scaled by 1/R. The reason for scaling is that the variances computed at a resolution in frequency greater than the harmonic resolution are not independent (as are the harmonic variances). The dependence is taken into account by reducing the bandwidth associated with each high resolution spectrum variance by 1/R.

Using padding for FFT purposes, with the consequent increase in spectral resolution, the expression for the periodogram variance is

$$
S_r' \times S_r'^* = A_r'^2 + B_r'^2 = \frac{1}{(RN)^2} \left| \sum_{n=0}^{N-1} x_n \exp(-i2\pi rn/RN) \right|^2,
$$

$$
r = -[(RN-1)/2], \ldots, 0, \ldots, [RN/2]. \tag{1.76}
$$

However, as mentioned earlier, to match the variances associated with high resolution data set A, the Fourier coefficients in data set B have to be multiplied by R, or, what is the same, the variances in Equation 1.76 have to be multiplied by R^2. Thus,

$$
A_r'^2 + B_r'^2 = R^2 A_r'^2 + R^2 B_r'^2. \tag{1.77}
$$

It follows that for high-resolution data set A, the total variance is given by

$$\frac{1}{R} \sum_{r=-[(RN-1)/2]}^{[RN/2]} (A_r'^2 + B_r'^2) \tag{1.78}$$

and, for data set B, by

$$\frac{1}{R} \sum_{r=-[(RN-1)/2]}^{[RN/2]} R^2 \left(A_r'^2 + B_r'^2\right) = R \sum_{r=-[(RN-1)/2]}^{[RN/2]} (A_r'^2 + B_r'^2). \tag{1.79}$$

If the variances at the nonharmonic frequencies in Equation 1.77 are treated as random variables, they have the same asymptotic (as N tends to infinity) mean, variance, and distribution as those at the harmonic frequencies (Koopmans, 1974, pp. 261–265).

The consequences of padding a time series with zeroes to accommodate analysis with an FFT can be illustrated with a typical example in which the number of data does not match the requirements of the FFT that is to be used. Consider spatially averaged sea surface temperature (SST) in an area bounded by 10° south latitude on its southern edge, the equator on its northern edge, and 80 and 90° west longitude on its eastern and western edges, respectively. This area comprises Niño Regions 1 and 2, which are often used as indicator regions of the current state of the El Niño Southern Oscillation (ENSO). Figure 1.46 shows Niño Region 1 + 2 monthly SST anomalies for the 30-year period of 1981–2010. These values were derived by subtracting the 1981–2010 monthly means from each month's actual SST in a manner identical to that used in Section 1.4.6.2. The original monthly values of SST were obtained from

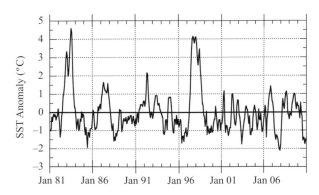

Figure 1.46   Mean monthly SST (sea surface temperature) anomalies in Niño Region 1 + 2 for the 30-year period of 1981–2010.

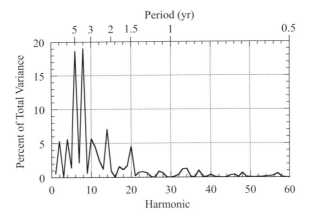

Figure 1.47 Periodogram to harmonic 60 of SST anomalies in Figure 1.46. Total anomaly variance is $1.440\,°C^2$.

the National Weather Service Climate Prediction Center (http://www.cpc.ncep. noaa.gov/data/indices/).

   If one were not concerned with computational speed, the SST anomaly data could be analyzed without implementation of an FFT by using standard Fourier analysis techniques, as in Appendix 1.A, where the number of data is $N = 360$. The first 60 harmonics of the periodogram resulting from such an analysis are presented in Figure 1.47, where these harmonics explain over 96% of the total variance in the time series. Notice that harmonics 30 and 60 necessarily have zero variance, since the 30-year mean was removed from the data. There is a concentration of variance at harmonics 20 and lower, corresponding to periods of 1.5 years and longer, and of particular interest are the peak variances that occur at periods of 3.75 and 5.0 years (harmonics 8 and 6, respectively). The periodogram provides an informative depiction of the cyclic nature of El Niño and La Niña.

   Suppose, however, that many thousands of such analyses needed to be performed as rapidly as possible. The computational efficiency of FFTs then becomes necessary, and it may be the case that the FFT requires $N = 2^k$ data points as previously described. By augmenting the 30 years of monthly data with 152 zeroes, we obtain a time series that is 512 (2^9) data points in length. An FFT was used to compute the folded version (positive harmonics only) of the Fourier coefficients in Equation 1.73 and the folded version of the periodogram variances in Equation 1.76 where $RN = 512$. The ratios (in percentage) of the individual variances to their sum are shown in Figure 1.48 for the first 85 harmonics to account for the inherently narrower bandwidth of the analysis.

   In comparing Figures 1.47 and 1.48 to each other, we can make the general statement that they are different for two reasons. One is that the period (in years)

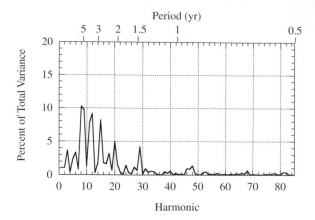

Figure 1.48   Periodogram to harmonic 85 of SST anomalies in Figure 1.46 after padding with 152 zeroes. Total anomaly variance is $1.440\,°C^2$.

versus harmonic relation is different in each figure; the other is that the bandwidth or width of the spectrum window associated with the variance estimates in each figure is different. To explain these differences, consider an example. As mentioned earlier, Figure 1.47 shows two strong peaks: the first at a period of 5 years (harmonic 6) and the second at a period of 3.75 years (harmonic 8). In Figure 1.48, the periodogram of the padded data, the variance in the first peak lies between harmonic 8, corresponding to a period of 5.333 years, and harmonic 9, corresponding to a period of 4.741 years. The spectrum windows centered at harmonics 8 and 9 transfer most of the variance in the first peak in Figure 1.47 to these two harmonics. Because the location of the first peak is approximately midway between the surrounding harmonics 8 and 9 in Figure 1.48, the variances there are quite similar.

An analogous situation occurs with the second peak in variance in Figure 1.47. This peak, at a period of 3.75 years (harmonic 8), lies between harmonics 11 (period of 3.879 years) and 12 (period of 3.556 years). Again, the spectrum windows centered at these harmonics redistribute the variance that lies between them in Figure 1.47 to these harmonics, as shown in Figure 1.48.

The opposite situation occurs in Figure 1.48 at harmonic 15 and period 2.844 years. The peak at this harmonic falls between harmonics 10 (period of 3 years) and 11 (period of 2.727 years) in Figure 1.47. In this case, a redistribution of a peak in variance in the padded spectrum occurs in the unpadded spectrum. In any periodogram, variance that is intrinsically located between harmonics is distributed to surrounding harmonics by a spectrum window centered at each harmonic, as we learned in Section 1.5.2. In addition, the two periodograms of the SST anomaly data illustrate the effect of data length on finding periodicities, a subject discussed in Section 1.5.4.

Appendix 1.A   Subroutine foranx

```
subroutine foranx (s, n, var, nf, tvar, fr1, fr2, iprnt)
dimension s(1), a(400), b(400), pvar(400), phi(400), var(400),
freq(400)
c*****************************************************************
c
c     This subroutine performs a fast Fourier analysis of an even
      number of data points at as
c     many frequencies as desired. The frequency span is between 0.0
      and 0.5 cy/data interval,
c     inclusive. The algorithm used is that given at the end of
      Chapter 9 of Spectral Analysis
c     (Jenkins & Watts, 1968, Holden-Day, San Francisco, 525 pp.)
c         * Input *
c  s  input data array.
c  n  length of s. n is an even number.
c  nf  number of frequencies (including zero) at which variance
      is to be computed.
c     nf.gt.n/2 and is an odd number.
c     nf = n/2 + 1 for standard periodogram.
c  fr1 frequency at which printing begins.
c  fr2 frequency at which printing ends.
c      fr1 and fr2 are less than or equal to 0.5 and fr1 < fr2.
c          * Output *
c     var the array of spectrum variances at the nf frequencies.
c     tvar the total variance in the data.
c           * Other *
c     iprnt user supplied output device unit number.
c
c     the synthesis form is x(t) = a*cos(wt) + b*sin(wt)
c          = c*cos(wt - phi) phi = phase angle in program
c
c*****************************************************************
c     ** set constants, variance scale factor, and frequency array **
c
      data pi /3.1415926536/
      anf = nf
      an = n
      rddg = 180.0/pi
      frfac = an/(2.0*anf - 2.0)
      bb = 0.5/(anf - 1.0)
      do 10 i = 1, nf
      aa = i
   10 freq(i) = (aa - 1.0)*bb
```

```
c
c      ** get Fourier coefficients at 0 and Nyquist frequencies **
c
       a(1) = 0.0
       b(1) = 0.0
       a(nf) = 0.0
       b(nf) = 0.0
       e = n
       tvar = 0.0
       do 20 i = 1, n
       c = i - 1
       r = s(i)/e
       a(1) = a(1) + r
 20    a(nf) = a(nf) + r*cos(c*pi)
c
c      ** get variance in data, start accumulation of variance in
       spectrum **
c
       do 30 i = 1, n
       s(i) = s(i) - a(1)
 30    tvar = tvar + s(i)**2/e
       var(nf) = a(nf)**2
       pvar(nf) = var(nf)*100.0*frfac/tvar
       var(1) = 0.0
       pvar(1) = 0.0
       wvar = var(nf)*frfac
       g = n/2
       phi(1) = 0.0
       phi(nf) = 0.0
       do 40 i = 1, n
 40    s(i) = s(i)/g
c
c      ** J & W algorithm **
c
       nfm1 = nf - 1
       do 50 j = 2, nfm1
       ang = 2.0*pi*freq(j)
       co = cos(ang)
       si = sin(ang)
       v0 = 0.0
       v1 = 0.0
       z0 = 0.0
       z1 = 0.0
       do 60 i = 2, n
       ii = n - i + 2
```

```
          v2 = 2.0*co*v1 - v0 + s(ii)
          z2 = 2.0*co*z1 - z0 + s(ii)
          v0 = v1
          v1 = v2
          z0 = z1
          z1 = z2
   60     continue
          a(j) = s(1) + v1*co - vo
          b(j) = z1*si
          var(j) = (a(j)**2 + b(j)**2)/2.0
          pvar(j) = var(j)*frfac*100.0/tvar
          wvar = wvar + var(j)*frfac
   50     phi(j) = atan2(b(j), a(j))*rddg
   C
   C      **print results **
   C
          kj = 0
          ih1f = 1
          ih2f = nf
          if(fr1.lt.0.0.or.fr1.gt.0.5) go to 111
          if(fr2.lt.fr1.or.fr2.gt.0.5) go to 111
          if(fr1.eq.0.0.and.fr2.eq.0.5) go to 109
          ih1f = 2.0*fr1*(anf - 1.0) + 1.01
          ih2f = 2.0*fr2*(anf - 1.0) + 1.01
   C
  109     do 70 j = ih1f, ih2f
          wn = freq(j)*an
          kj = kj + 1
          if(((kj - 1)/25)*25.eq.(kj - 1)) write(iprnt, 102)
   70     write(iprnt,103) wn, freq(j), a(j), b(j), var(j), pvar(j),
          phi(j)
          write(iprnt,104) tvar, wvar
  102     format(//9x, 'har-', 6x, 'freq', 7x, 'cosine', 9x, 'sine',
         *10x, 'line', 7x, 'percent of', 5x, 'phase', /8x, 'monic',
         *5x, 'cy/di', 2(3x,'coefficient'), 5x, 'variance', 6x,
          'total var',*5x, 'angle', /)
  103     format(5x, f8.3, 4x, f6.3, 4(2x,g12.5), 4x, f6.1)
  104     format(///21x, 'variance in data set', g12.5//8x,
         *'variance explained by periodogram', g12.5)
          go to 99
  111      write(iprnt, 112)
  112     format(//, 10x, 'fr1 or fr2 or both out of range')
   99     return

          end
```

Appendix 1.B   Sum of complex exponentials

Let

$$Q = \sum_{n=a}^{b} \exp(i\omega n) = e^{i\omega a} + e^{i\omega(a+1)} + \cdots + e^{i\omega b} \qquad (1.B.1)$$

where a and b are integers and b > a. Multiply Equation 1.B.1 by $e^{i\omega}$ to get

$$e^{i\omega}Q = e^{i\omega(a+1)} + e^{i\omega(a+2)} + \ldots + e^{i\omega(b+1)}. \qquad (1.B.2)$$

Subtract Equation 1.B.2 from Equation 1.B.1 to obtain

$$Q = \frac{\left(e^{i\omega a} - e^{i\omega(b+1)}\right)}{\left(1 - e^{i\omega}\right)}. \qquad (1.B.3)$$

Now multiply the numerator and denominator of Equation 1.B.3 by $\exp(-i\omega/2)$. Then successively withdraw $\exp(ia\omega/2)$ and $\exp(ib\omega/2)$. The result is

$$Q = \sum_{n=a}^{b} \exp(i\omega n)$$

$$= \exp[i(a+b)\omega/2] \left[\frac{e^{i\omega(b-a+1)/2} - e^{-i\omega(b-a+1)/2}}{e^{i\omega/2} - e^{-i\omega/2}} \right]$$

which, using Euler's formula, reduces to

$$Q = \exp[i(a+b)\omega/2] \frac{\sin[\omega(b-a+1)/2]}{\sin(\omega/2)}. \qquad (1.B.4)$$

In application of Equations 1.B.3 and 1.B.4 it is important to test $\sin(\omega/2)$ to verify that it is not zero for any values of the argument. If $\sin(\omega/2)$ is zero, then l'Hopital's rule can be applied to these equations to obtain a determinate form. Equation 1.B.1 can be used, also.

Appendix 1.C   Distribution of harmonic variances

The purpose of this appendix is to develop relationships for the statistical distribution of the harmonic variances. Because the chi-square (χ^2) distribution plays a prominent role in the development that follows, it is important to be familiar with its properties. We begin with the forms for the Fourier cosine and sine amplitudes given in Table 1.1, now treated as random variables, for an even number of data N, namely

$$A_m = \frac{2}{N}\sum_{n=0}^{N-1} X_n \cos\frac{2\pi mn}{N} \qquad (1.C.1a)$$

and

$$B_m = \frac{2}{N}\sum_{n=0}^{N-1} X_n \sin\frac{2\pi mn}{N}, \qquad m = \left[0, \frac{N}{2}\right]. \qquad (1.C.1b)$$

Making use of linear expectation operator E and assuming a purely random process (white noise) represented by random variable X_n with $E[X_n]=0$ so that $E[A_m]=E[B_m]=0$, we obtain:

$$\mathrm{Var}[A_m] = E[A_m^2]$$

$$= \frac{4}{N^2}\left\{ E[X_0X_0]\cos\frac{2\pi m0}{N}\cos\frac{2\pi m0}{N} + E[X_0X_1]\cos\frac{2\pi m0}{N}\cos\frac{2\pi m1}{N}\right.$$

$$+\cdots+ E[X_0X_{N-1}]\cos\frac{2\pi m0}{N}\cos\frac{2\pi m(N-1)}{N} + E[X_1X_0]\cos\frac{2\pi m1}{N}\cos\frac{2\pi m0}{N}$$

$$+E[X_1X_1]\cos\frac{2\pi m1}{N}\cos\frac{2\pi m1}{N}+\cdots+ E[X_1X_{N-1}]\cos\frac{2\pi m1}{N}\cos\frac{2\pi m(N-1)}{N}$$

$$+\cdots+ E[X_{N-1}X_0]\cos\frac{2\pi m(N-1)}{N}\cos\frac{2\pi m0}{N}$$

$$+E[X_{N-1}X_1]\cos\frac{2\pi m(N-1)}{N}\cos\frac{2\pi m1}{N}$$

$$\left.+\cdots+E[X_{N-1}X_{N-1}]\cos\frac{2\pi m(N-1)}{N}\cos\frac{2\pi m(N-1)}{N}\right\}. \qquad (1.C.2)$$

The expectation $E[X_iX_j]=0$ for $i\neq j$ because the random variables are uncorrelated; similarly, $E[X_iX_j]=\sigma_X^2$ for $i=j$ because the random variables are completely correlated. The latter relation follows from Equation 1.18 and noting from above that $E[X_n]=0$. Therefore, Equation 1.C.2 becomes

$$
\text{Var}[A_m] = \begin{cases} \dfrac{4}{N^2}\sigma_X^2\displaystyle\sum_{n=0}^{N-1}\cos^2\dfrac{2\pi mn}{N} = \dfrac{4}{N^2}\sigma_X^2\dfrac{N}{2} = \dfrac{2}{N}\sigma_X^2, & m = \left[1, \dfrac{N}{2}-1\right] \\[3mm] \dfrac{1}{N^2}\sigma_X^2\displaystyle\sum_{n=0}^{N-1}\cos^2\dfrac{2\pi mn}{N} = \dfrac{1}{N^2}\sigma_X^2 N = \dfrac{1}{N}\sigma_X^2, & m = 0, \dfrac{N}{2} \end{cases} \tag{1.C.3}
$$

and for the sine coefficients

$$
\text{Var}[B_m] = \begin{cases} \dfrac{4}{N^2}\sigma_X^2\displaystyle\sum_{n=0}^{N-1}\sin^2\dfrac{2\pi mn}{N} = \dfrac{4}{N^2}\sigma_X^2\dfrac{N}{2} = \dfrac{2}{N}\sigma_X^2, & m = \left[1, \dfrac{N}{2}-1\right] \\[3mm] \dfrac{1}{N^2}\sigma_X^2\displaystyle\sum_{n=0}^{N-1}\sin^2\dfrac{2\pi mn}{N} = 0, & m = 0, \dfrac{N}{2}. \end{cases} \tag{1.C.4}
$$

The sums of the cosine-squared and sine-squared terms can be determined from Equations 1.4 and 1.5.

The covariance between the coefficients at different harmonics may be calculated in a similar manner. For $m \neq k$,

$$
\text{Cov}[A_m, A_k] = \dfrac{4}{N^2}\sigma_X^2\sum_{n=0}^{N-1}\cos\dfrac{2\pi mn}{N}\cos\dfrac{2\pi kn}{N}
$$
$$
= 0 \tag{1.C.5}
$$

since the cosine terms are orthogonal to each other, as demonstrated in the derivation of Equation 1.8. Similarly, for the sine coefficients

$$
\text{Cov}[B_m, B_k] = 0. \tag{1.C.6}
$$

Lastly, for all m, k,

$$
\text{Cov}[A_m, B_k] = 0 \tag{1.C.7}
$$

because over their length, any integer number of sine waves is orthogonal to any integer number of cosine waves.

Now assume that each rv X_n from our white noise process has a normal distribution with population mean zero and population variance σ_X^2. Since random variables A_m and B_m are linear functions of normal random variables from Equation 1.C.1, they also are normally distributed. From statistical theory, the square of a normal random variable with zero mean and unit variance (i.e., a standard normal variable) is distributed as a chi-square variable with one degree of freedom. Thus, if we square the Fourier coefficients and standardize them by dividing by their variance, we have

$$\frac{A_m^2}{\text{Var}[A_m]} = \frac{A_m^2}{\sigma_X^2/(N/2)} \Rightarrow \chi_1^2, \quad m = \left[1, \frac{N}{2} - 1\right] \tag{1.C.8}$$

$$\frac{B_m^2}{\text{Var}[B_m]} = \frac{B_m^2}{\sigma_X^2/(N/2)} \Rightarrow \chi_1^2, \quad m = \left[1, \frac{N}{2} - 1\right] \tag{1.C.9}$$

and

$$\frac{A_m^2}{\text{Var}[A_m]} = \frac{A_m^2}{\sigma_X^2/N} \Rightarrow \chi_1^2, \quad m = 0, \frac{N}{2} \tag{1.C.10}$$

in which the arrow indicates "is distributed as." Notice that no equation comparable to Equation 1.C.10 is given for B_m when $m = 0$, $N/2$; the reason is that B_m is always zero for these two values of m. There is another relevant relationship involving χ^2 variables: the sum of any number of mutually independent χ^2 variables whose degrees of freedom sum to v is itself a χ^2 variable with v degrees of freedom; that is,

$$\chi_{v_1}^2 + \chi_{v_2}^2 + \cdots + \chi_{v_k}^2 = \chi_v^2 \tag{1.C.11}$$

where $v = v_1 + v_2 + \cdots + v_k$. Thus, dividing Equations 1.C.8 and 1.C.9 by two and then summing yields

$$\frac{\frac{A_m^2 + B_m^2}{2}}{\sigma_X^2/(N/2)} \Rightarrow \frac{\chi_2^2}{2}, \quad m = \left[1, \frac{N}{2} - 1\right]. \tag{1.C.12}$$

The reason for dividing by two is to match the expression for variance at a harmonic given in Table 1.1. The denominators in Equations 1.C.12 and 1.C.10 distribute the population variance σ_X^2 among the harmonic frequencies in such a way that the variance at the interior harmonics is uniform and twice the value at the frequency origin ($m = 0$) and the highest frequency ($m = N/2$). The variance at $m = 0$ is the variance of the sample mean (i.e., the mean of a realization) about the population mean, the latter value of which is zero in this development.

Now simplify the notation by letting

$$C(f_m) = \frac{A_m^2 + B_m^2}{2}, \quad m = \left[1, \frac{N}{2} - 1\right] \tag{1.C.13a}$$

and

$$C(f_m) = A_m^2, \quad m = 0, \frac{N}{2} \tag{1.C.13b}$$

where $f_m = m/N\Delta t$ is the harmonic frequency for harmonic m. Next, replace the white noise variance $\sigma_X^2/(N/2)$ at the interior harmonics and σ_X^2/N at the two exterior harmonics, by $\Gamma(f_m)$ in Equations 1.C.12 and 1.C.10. Taking their expectations, and noting that $E[\chi_v^2] = v$, yields

$$E\left[\frac{C(f_m)}{\Gamma(f_m)}\right] = E\left[\frac{\chi_2^2}{2}\right] = 1, \quad m = \left[1, \frac{N}{2} - 1\right] \tag{1.C.14}$$

and

$$E\left[\frac{C(f_m)}{\Gamma(f_m)}\right] = E[\chi_1^2] = 1, \quad m = 0, \frac{N}{2} \tag{1.C.15}$$

with the result that

$$E[C(f_m)] = \Gamma(f_m), \quad m = \left[0, \frac{N}{2}\right]. \tag{1.C.16}$$

We now introduce the term *estimator*. An estimator is a random variable used to estimate a population parameter. For example, in Equation 1.C.13, spectrum estimator $C(f_m)$, as an appropriate function of the Fourier coefficients, is used to estimate the population variance at frequency f_m. Equation 1.C.16 shows that $C(f_m)$ is an unbiased estimator of the white noise variance at the harmonic frequencies because its expected value is equal to the population variance $\Gamma(f_m)$. If the expected value were something other than $\Gamma(f_m)$, $C(f_m)$ would be a biased estimator. It is usually desirable that an estimator be unbiased. However, if the calculation of an unbiased estimator requires information that is otherwise unavailable, or if repeated calculations are needed that consume significant computation time, it may be more advantageous to employ a biased estimator.

Since $Var[\chi_v^2] = 2v$, we have, following Equations 1.C.14 and 1.C.15,

$$Var[C(f_m)] = \Gamma^2(f_m), \quad m = \left[1, \frac{N}{2} - 1\right] \tag{1.C.17}$$

and

$$Var[C(f_m)] = 2\Gamma^2(f_m), \quad m = 0, \frac{N}{2} \tag{1.C.18}$$

showing that the variance of the estimator is uniform at the interior harmonics and twice that value at the exterior harmonics (note the definitions of $\Gamma(f_m)$). That we are dealing with the variance of harmonic variances simply means that each harmonic variance $C(f_m)$ is itself a random variable and thus has a probability distribution function based on an infinity of realizations. Equations 1.C.16–1.C.18 are the expressions for the mean and variance of the probability distribution function.

It is significant that the variances of the harmonic variances are independent of sample size. The collection of additional data does not increase the stability of the estimator. That this is the case is not unexpected because as the length, N, of the time series increases, the number of Fourier harmonics increases accordingly and the separation between them, that is, the bandwidth or frequency averaging distance associated with each harmonic, decreases. The number of data (degrees of freedom for white noise) consumed in a variance estimate remains the same. To effect increased stability of the spectrum estimator requires some form of spectrum averaging.

In the case of N odd, the analysis parallels that above, except that the highest harmonic is $(N-1)/2$. To get $\Gamma(f_m)$ at all harmonics except the frequency origin divide the population variance by N/2; at $m = 0$ divide σ_X^2 by N.

The above derivations have been done under the assumption that the population mean is known, and in this case equal to zero. The derivation could have been done with a known nonzero mean, but the procedure is more tedious. More generally, the population mean is unknown and the total variance in a given time series is taken with respect to the sample mean. If the time series is hypothesized to be a realization of white noise (with mean unknown), the total variance is similarly distributed as above but without any variance contribution at $m = 0$. This is because the total variance must be perforce computed about the sample mean.

For the case of an even number of data, the estimate of the total variance, $\hat{\sigma}_X^2$, is divided by $(N-1)/2$ to obtain white noise variances at the interior harmonics and by $N-1$ to obtain the white noise variance at the highest harmonic, N/2. There is no contribution of variance at $m = 0$. In the case of an odd number of data, the total variance is divided by $(N-1)/2$ to obtain estimates of the harmonic white noise variances and, again, there is no contribution of variance at $m = 0$.

In summary, for N even and the mean of the white noise process known, variances at the interior harmonics have a distribution proportional to $\chi_2^2/2$. Variances at the two exterior harmonics (0 and N/2) have a distribution proportional to $\chi_1^2/1$. For N odd and the population mean known, the distributions of variance at all harmonics are proportional to $\chi_2^2/2$, except at the 0-th harmonic where the distribution is proportional to $\chi_1^2/1$. When the population mean is unknown, the variances have similar distributions except that no variance is generated at the origin.

Appendix 1.D　Derivation of Equation 1.42

The problem is to find the variance at any harmonic frequency when the input is at a nonharmonic frequency. Consider the general input sinusoid $a\cos(\omega n - \phi)$ and take its Fourier transform. From Equation 1.63 for a two-sided spectrum,

$$S'_m = A'_m - iB'_m = \frac{1}{N}\sum_{n=0}^{N-1} a\cos(\omega n - \phi)\exp(-i\omega_m n)$$

$$m = -[(N-1)/2], \ldots, 0, \ldots, [N/2] \qquad (1.D.1)$$

where m is harmonic number and $\omega_m = 2\pi m/N$ is angular frequency. It is assumed that the time step $\Delta t = 1$.

Using Euler's formula, the input sinusoid can be put in complex exponential form such that:

$$(A'_m - iB'_m)\frac{2N}{a} = \exp(-i\phi)\sum_{n=0}^{N-1}\exp[i(\omega - \omega_m)n] + \exp(i\phi)\sum_{n=0}^{N-1}\exp[-i(\omega + \omega_m)n].$$

$$(1.D.2)$$

From Equation 1.B.4,

$$(A'_m - iB'_m)\frac{2N}{a} = \exp\{i[(N-1)(\omega - \omega_m)/2 - \phi]\}\frac{\sin[N(\omega - \omega_m)/2]}{\sin[(\omega - \omega_m)/2]}$$

$$+ \exp\{-i[(N-1)(\omega + \omega_m)/2 - \phi]\}\frac{\sin[N(\omega + \omega_m)/2]}{\sin[(\omega + \omega_m)/2]}.$$

$$(1.D.3)$$

We can make use of Euler's formula, again, to rewrite the exponential terms of Equation 1.D.3. Equating the real portions of the resulting equation allows us to solve for A'_m, and, similarly, equating the imaginary portions yields B'_m, so that

$$A'_m = \frac{a}{2}\left\{\cos\left[(N-1)\left(\frac{\omega + \omega_m}{2}\right) - \phi\right]\frac{\sin N[(\omega + \omega_m)/2]}{N\sin[(\omega + \omega_m)/2]}\right.$$

$$\left. + \cos\left[(N-1)\left(\frac{\omega - \omega_m}{2}\right) - \phi\right]\frac{\sin[N(\omega - \omega_m)/2]}{N\sin[(\omega - \omega_m)/2]}\right\} \qquad (1.D.4)$$

and

$$
B'_m = \frac{a}{2} \left\{ \sin\left[(N-1)\left(\frac{\omega+\omega_m}{2}\right) - \phi\right] \frac{\sin[N(\omega+\omega_m)/2]}{N\sin[(\omega+\omega_m)/2]} \right.
$$

$$
\left. -\sin\left[(N-1)\left(\frac{\omega-\omega_m}{2}\right) - \phi\right] \frac{\sin N[(\omega-\omega_m)/2]}{N\sin[(\omega-\omega_m)/2]} \right\}. \tag{1.D.5}
$$

These results apply for N even or odd and to a two-sided spectrum. In reference to Equation 1.42, where N is even and the periodogram is one sided, the A'_m and B'_m above have to be doubled except at $m=0$, $N/2$. Thus the variance at positive harmonic m is

$$
S_m^2(\omega) = \left[(2A_m)^2 + (2B_m)^2\right]/2 = 2A_m^2 + 2B_m^2
$$

$$
= \frac{a^2}{2} \left\{ \frac{\sin^2[N(\omega+\omega_m)/2]}{N^2\sin^2[(\omega+\omega_m)/2]} + \frac{\sin^2[N(\omega-\omega_m)/2]}{N^2\sin^2[(\omega-\omega_m)/2]} \right.
$$

$$
+ 2\cos[(N-1)\omega - 2\phi]\frac{\sin[N(\omega+\omega_m)/2]}{N\sin[(\omega+\omega_m)/2]}
$$

$$
\left. \times \frac{\sin[N(\omega-\omega_m)/2]}{N\sin[(\omega-\omega_m)/2]} \right\}, \qquad m\neq 0, \frac{N}{2} \tag{1.D.6}
$$

which is Equation 1.42.

Problems

1 On graph paper, carefully sketch at least one complete cycle of the sinusoid given by

$$
y(t) = 1 - 2\cos(0.5\pi t + \pi/4)
$$

starting at $t=-1$. (Suggestion: First find the period and location of the maximum or minimum of the cosine term alone without the phase angle. Then adjust the plot to take into account phase and vertical displacement.)

2 The figure below shows a sinusoid that is digitally sampled according to

$$x_k = a + c\cos\left(\frac{2\pi mk}{N} - \phi_m\right), \quad k = 0, \pm 1, \pm 2, \ldots$$

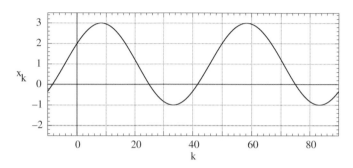

From the above figure determine:

(a) $a = $ _____

(b) $c = $ _____

(c) an appropriate $m = $ _____ for an appropriate $N = $ _____

(d) $\phi_m = $ _____ degrees

3 Use Appendix 1.B to show that

$$\sum_{n=0}^{N-1} \sin^2\left(\frac{2\pi kn}{N}\right) = \frac{N}{2}$$

where N is an even integer and $0 \le k \le N/2$.

4 A time series of length $N\Delta t$ where $N = 50$ is obtained. It then is discovered that the last half of the series, $25\Delta t$, is a repeat of the first $25\Delta t$. How does the variance of the time series of length $50\Delta t$ compare with the variance of the time series of length $25\Delta t$?

5 Manual Fourier Analysis: use only paper, pencil, and a nonprogrammable hand-held calculator.

The data below are 30-year normal monthly precipitation values for 1971–2000 at San Francisco International Airport (SFO AP), California (37.62 N, 122.40 W) and Oklahoma City Will Rogers Airport (OKC AP), Oklahoma (35.38 N, 97.60 W).

(a) Plot the data on separate graphs and comment on differences you observe between the time series. Can you provide a meteorological explanation for the differences in annual total precipitation regimes?

(b) Choose one of the time series and perform a Fourier analysis of the data sufficient to detect both amplitudes and phase angles of the significant harmonics present (i.e., find enough harmonics to explain at least 95% of the variance in the data).

(c) On a separate graph, plot the significant waves in (b) in the form of the amplitude and phase representation discussed in Section 1.2.4.

(d) Plot the sum of the significant waves (plus the mean) on the time series graph in (a).

(e) What percentage of the variance of the observed series does each harmonic explain?

(f) Compare the observed variance (that of the data set itself) with the explained variance to obtain residual variance.

Month	San Francisco International Airport Precipitation (mm)	Oklahoma City Will Rogers Airport Precipitation (mm)
January	113.0	32.5
February	101.9	39.6
March	82.8	73.7
April	30.0	76.2
May	9.7	138.2
June	2.8	117.6
July	0.8	74.7
August	1.8	63.0
September	5.1	101.1
October	26.4	92.5
November	63.2	53.6
December	73.4	48.0

6 Fourier Analysis Using a Computer Program

In this problem we use the paradrop days data in Table 1.3 that were discussed in Section 1.3.2.

(a) Write a computer program that will find the cosine amplitudes, sine amplitudes and phase angles for the largest harmonics that explain at least 95% of the variance.

(b) Convert the phase angles into actual times of the maximum amplitude for the various harmonics.

(c) Plot (1) the original data, (2) each of the largest harmonics, the sum of which explains at least 95% of the variance, and (3) the sum of these harmonics, all on one graph. Compare your results with Figure 1.14.

(d) Can you attach any physical meaning to the individual harmonics? You might consider the typical cycle of daily wind, for example. Does it have a sinusoidal shape?

7 Recall that the variance of rv X is given by

$$Var[X] = \int_{-\infty}^{\infty} (x - \mu)^2 f(x)dx.$$

Let rv X have a uniform probability density function $f(x)$ between a and b and zero elsewhere. If $b - a = 1$, find the variance of rv X for this rectangular distribution.

8 The observed variance in a periodogram at harmonic k is $8\,°C^2$. The goal is to find the limits of the 95% *a priori* confidence interval for the population variance $\Gamma(f_k)$ at harmonic k. Assume that

$$\frac{C(f_k)}{\Gamma(f_k)} = \chi_2^2/2$$

(a) Write down the appropriate probability statement(s) of the form $Pr\{\_\_\} = \_\_$ for the confidence limits on the population variance

(b) What are the upper and lower limits of the 95% *a priori* confidence interval? Recall that

$$f_{\chi_2^2}(x) = \frac{1}{2}e^{-x/2}$$

9 The observed variance in a periodogram of a time series with $N = 41$ data is found to be $12\,m^2$. The null hypothesis is made that the sample of data comes from a white noise process. Find the limits of the 95% *a posteriori* confidence interval for the observed variances at the harmonic frequencies.

10 Consider a time series comprising $N = 51$ data with variance $= 40\,Pa^2$. The null hypothesis H_0 is made that the realization is from a data population that is white noise. A periodogram of the time series is calculated and the largest value in the periodogram is $10.45\,Pa^2$ and the smallest is $0.0065\,Pa^2$. Show whether H_0 will be rejected or not rejected.

11 If a signal can be described by $x_n = A \sin(2\pi f n \Delta t)$ in which $\Delta t = 0.1$ s and $f = 12$ Hz, at which frequencies in the principal part of the complete aliased spectrum will the variance be observed and what will be the variance at each frequency?

12 Suppose you have a data set that comprises 100 values of wind speed in which the sampling interval is two seconds. Unbeknownst to you, there was a strong sinusoid with period 1.6 seconds introduced into the analog signal (i.e., before digitization) because of a defective electronic component. A periodogram analysis of the data set is performed.

(a) What is the Nyquist frequency in Hz?

(b) At what positive frequency (in Hz) in the principal part of the aliased spectrum will the erroneous variance occur?

(c) What is the corresponding harmonic number for the frequency found in (b)?

(d) What can be done or what should have been done to eliminate the unwanted signal from appearing in the periodogram? Explain.

13 An analog temperature signal is sampled once every second. The number of data collected is 40. Unfortunately, a nearby transmitter has added an unwanted frequency of 1.125 Hz.

(a) At what *frequencies* (Hz) in the principal part of the (two-sided) aliased spectrum will the unwanted variance appear?

(b) What are the corresponding *harmonics* in the principal part of the aliased spectrum at which the variances occur?

14 Consider a stagecoach scene in a motion picture (e.g., *How the West Was Won*). The wheels of the stagecoach have a radius $r = 0.6$ m and each has eight spokes. Assume the camera shutter speed is 24 frames per second.

Plot the perceived (which may be the actual) angular speed (radians/second) of one of the wheels versus the speed of the stagecoach as it increases from 0 m/s to the speed at which the wheels are perceived to be stationary, that is, not rotating. (Hint: Sketch an eight-spoke wheel, write down the equation for the stagecoach speed in terms of the angular speed of a wheel, then adapt it to the conditions of the problem.)

15 Under certain conditions the spectrum window function of the form $[(\sin x)/x]^2$ can be used to estimate the variance at harmonic frequencies due to variance in the data at nonharmonic frequencies.

(a) What are the two primary conditions?

(b) Assume the conditions of (a) are met. Sketch at least one spectrum window centered at a harmonic and calculate the variance at harmonics $m-1$, m, and $m+1$ based on the figure below. The one-sided variance input shown in the figure is $10\,\mathrm{m^2\,s^{-2}}$ located midway between harmonics $m-1$ and m.

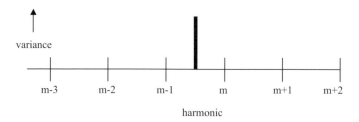

16 The objectives of this problem are to compare periodograms of hourly temperature for January and July, 2009 at Oklahoma City, OK, and determine whether the hourly temperatures in these months can be modeled as a white noise process after removal of the daily cycle.

Data

The data are available on the website http://www.wiley.com/go/duchon/ timeseriesanalysis. The filenames are OKC_200901_hrly_temp.xls and OKC_200907_hrly_temp.xls. The data are hourly temperatures in degrees Celsius for January and July 2009. The first column is the sequential hour count, the second column is the date, the third column is the time the temperature was observed in Central Standard Time, and the fourth column is the temperature. The only data needed to work this problem are the hourly temperatures in the fourth column.

(a) Plot the times series of hourly temperature for each month on separate sheets of paper, using the same size for all your plots. Show on each plot frontal passages, cloudy days, clear days, and any other meteorological events that you believe to be present.

(b) Use the Fourier Analysis computer program you designed in problem 6 or subroutine FORANX in Appendix 1.A to compute the periodogram of the 744 points for each month. Plot the $\log_{10}$ variance (or variance on a $\log_{10}$ axis) versus frequency, period, or harmonic for all harmonics. On each plot show the total variance and bandwidth associated with each plot. Place the plots on separate pages.

(c) Compute and plot the *average daily cycle* of temperature for each month. Briefly discuss the principal differences between the two months and their causes.

(d) Remove the daily cycle from the original hourly data for each month to form two new time series (the "uncontaminated" data). Plot the two time series of "uncontaminated" data. Comment on the presence or absence of the daily cycle of temperature.

(e) Plot the periodograms of the uncontaminated hourly data, replacing the variance estimates at the harmonic frequencies of the daily cycle with the average of surrounding variances. On each plot show the total variance and the bandwidth associated with each estimate.

(f) Apply a white noise test to each periodogram in (e). Compute the *a priori* confidence limits and *a posteriori* confidence limits. Place them on the periodograms of variance in which the vertical axis is $\log_{10}$ variance. Do you accept or reject the white noise null hypothesis? If you reject the hypothesis that the sample comes from a population of white noise, what physical phenomenon or phenomena do you think led to its rejection?

References

Bloomfield, P. (1976) *Fourier Analysis of Time Series: An Introduction*. John Wiley & Sons, Inc., New York.

Box, G.E.P., and Jenkins, G.M. (1970) *Time Series Analysis Forecasting and Control*. Holden-Day, San Francisco, CA.

Crowley, K.D., Duchon, C.E., and Rhi, J. (1986) Climate record in varved sediments of the eocene Green River formation. *J. Geophys. Res.*, **91**(D8), 8637–8647.

Daubechies, I. (1992) *Ten Lectures on Wavelets*. CBMS-NSF Regional Conference Series on Applied Mathematics, No. 61. Society for Industrial and Applied Mathematics (SIAM), Philadelphia, PA.

Hoel, P.G. (1962) *Introduction to Mathematical Statistics*. John Wiley & Sons, Inc., New York.

Jenkins, G.M., and Watts, D.G. (1968) *Spectral Analysis and its Applications*. Holden-Day, San Francisco, CA.

Koopmans, L.H. (1974) *The Spectral Analysis of Time Series*. Academic Press, New York.

McPherson, R.A., Fiebrich, C.A., Crawford, K.C. *et al.* (2007) Statewide monitoring of the mesoscale environment: a technical update on the Oklahoma Mesonet. *J. Atmos. Oceanic Technol.*, **24**, 301–321.

Nyquist, H. (1928) Certain topics in telegraph transmission theory. *Trans. Am. Inst. Electr. Eng.*, **47**, 617–644.

Robinson, E. A. (1982) A historical perspective on spectrum estimation. *Proc. IEEE*, **70**(9), 885–907.

2

Linear systems

The analysis of linear systems is fundamentally the study of the connection between two time series, one mathematically or physically created from the other. The complete system comprises an input time series, an output time series, and a *physical system* or *mathematical system* that provides the linkage between the input and output. A simple example of a physical system is an ordinary liquid-in-glass thermometer: it converts a change in temperature of its surroundings to a change in length of the column of liquid due to the expansion or contraction of the liquid in the bulb or reservoir. The complete system includes the air temperature (input), the thermometer (physical system), and the temperature lines or etched markings on the thermometer (output). Another example of a complete system is the amplifier (first part of the physical system) in a stereo receiver that magnifies a weak electrical signal (input) to sufficient strength to drive a speaker (second part of the physical system) that produces sound waves (output). A mathematical system consists of a filter or set of weights applied to a time series in order to alter its character in a predictable way. A common example is a running mean, which, when applied to a varying input time series, reduces the magnitude of fluctuations in the output time series. In addition to the mathematical system, the complete system includes the unfiltered input time series and the filtered output time series. While these are examples of simple systems, other examples consist of multiple systems linked in series or parallel. Their study can be very demanding.

No matter how simple or complex, the systems studied in this chapter are linear. Linear systems are much easier to analyze than nonlinear systems because of superposition; this is discussed in Section 2.1.

Time Series Analysis in Meteorology and Climatology: An Introduction, First Edition. Claude Duchon and Robert Hale.
© 2012 John Wiley & Sons, Ltd. Published 2012 by John Wiley & Sons, Ltd.

2.1 Input–output relationships

The motivation for this section is the schematic diagram in Figure 2.1, in which the wind speed x(t) is being sensed by an anemometer (a device to measure the speed of moving air) whose electrical output is amplified and recorded as a digital signal y(t) in a data logger. The physical system comprises the sensor, a signal adapter, and data logger, and its purpose is to provide an output signal y(t) that faithfully reproduces the input signal x(t) within the limitations of the sensor. In the development that follows, the electronic components of the physical system are assumed to function perfectly, so that differences between the input and output signals are due only to the properties of the sensor. In short, there is no electronic noise in the physical system. From a mathematical viewpoint, input–output relations can be more easily understood if both input and output are treated as analog time series, which will be the approach used here.

Possible questions we might ask about a complete system are:

(1) What will be y(t) given x(t) and the properties of the sensor?

(2) What was x(t) given y(t) and the properties of the sensor?

(3) What are the properties of the sensor, given x(t) and y(t)?

In many situations the properties of the sensor can be described mathematically as a time invariant linear ordinary differential or integro-differential equation. By time invariant is meant that the coefficients in the equation are constant with time. The general solution to this type of equation and the answer to question (1) is

$$y(t) = \int_{-\infty}^{\infty} h(u)\, x(t-u)\, du \qquad (2.1)$$

or, equivalently,

$$y(t) = \int_{-\infty}^{\infty} x(u)\, h(t-u)\, du \qquad (2.2)$$

where h(u) is the *system function* responsible for converting x(t) into y(t) and mathematically explains what happens inside the box in Figure 2.1. Each of the above

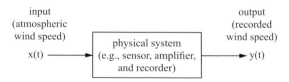

Figure 2.1   Schematic diagram of signal flow for a complete physical system.

expressions is called a *convolution integral* because of the location of variable t on the left side of the equation and inside the integral, where it is fixed with respect to integration. That Equations 2.1 and 2.2 are equivalent can be shown through a simple transformation of variables demonstrating that convolution obeys the commutative law.

What is meant by *linear* system? If we let $x_1(t)$ and $x_2(t)$ represent input signals, $y_1(t)$ and $y_2(t)$ output signals, and the signal flow by arrows, then in the two signal paths

$$x_1(t) \quad \rightarrow \quad [\text{system function}] \quad \rightarrow \quad y_1(t)$$

and

$$x_2(t) \quad \rightarrow \quad [\text{system function}] \quad \rightarrow \quad y_2(t)$$

the system is linear if

$$ax_1(t) + bx_2(t) \quad \rightarrow \quad [\text{system function}] \quad \rightarrow \quad ay_1(t) + by_2(t)$$

where a and b are arbitrary constants. If the last relationship is not true, the system is nonlinear. Concomitant with a linear system is the term *superposition*, which states that the output of a linear system having any number of inputs can be computed by determining the output of each input separately and then summing the individual outputs to obtain the total output as in the illustration above.

An example of a nonlinear system is the system equation

$$y(t) = bx^2(t) \tag{2.3}$$

where $x(t)$ is the input signal, $y(t)$ the output signal, and b is a constant. To test for superposition let the combined input at time t be $(x_1 + x_2)$; that is, there are two sources of input. Thus,

$$y \, (\text{combined input}) = b(x_1^2 + 2x_1x_2 + x_2^2). \tag{2.4}$$

The sum of the individual outputs, one from x_1, the other from x_2, and both passing through the system, is

$$y \, (\text{individual inputs}) = b(x_1^2 + x_2^2). \tag{2.5}$$

Outputs y for the combined and individual inputs are not the same; thus, super-position does not hold and the system is nonlinear. Of course, the system equation is obviously nonlinear because Equation 2.3 is a quadratic form.

The practical analysis of a system requires that it be *stable*. This means that if we have an input $x(t)$ which is bounded according to $|x(t)| \le k_1 < \infty$, where k_1

is a constant, the output $y(t)$ is also bounded according to $|y(t)| \le k_2 < \infty$, where k_2 is also a constant. It can be shown that a sufficient condition for system stability is that the integral of the magnitude of the system function is finite. Expressed mathematically,

$$\int_{-\infty}^{\infty} |h(u)|\, du \le k_3 < \infty$$

where k_3 is a third constant.

2.2 Evaluation of the convolution integral

Consider the convolution of two time-dependent functions $g_1(t)$ and $g_2(t)$, where $g_1(t)$ corresponds to $h(u)$ and $g_2(t-u)$ to $x(t-u)$ in Equation 2.1. The convolution operation is given formally by

$$g_3(t) = g_1(t)*g_2(t) = \int_{-\infty}^{\infty} g_1(u)\, g_2(t-u)\, du \qquad (2.6)$$

wherein the asterisk is often used as the convolution operator. The value of $g_3(t)$ for any particular time, t, is thus the area under the curve of the product of $g_1(u)$ and $g_2(t-u)$ over all time u. In addition, the arguments of g_1 and g_2 inside the integral sign are interchangeable. To understand the convolution technique, it is useful to visualize or sketch the relationship between $g_1(u)$ and $g_2(t-u)$ as time t changes. We do this in the next section for some simple functions, and for a first-order linear system in the subsequent section. Some of these illustrations are similar to convolution figures in Cooper and McGillem (1999, 1967).

2.2.1 Interpretation

There is no difficulty understanding $g_1(u)$ in Equation 2.6 because it is the same function as $g_1(t)$, except for a change in argument notation from t to u. However, understanding the function $g_2(t-u)$ requires some thought. Mathematically, $g_2(t-u)$ is a combination of *reflection* and *translation* of the original function $g_2(u)$. The process can be visualized through examples.

Consider the function $g_2(u)$ shown in Figure 2.2a. The function $g_2(-u)$ shown in Figure 2.2b is simply a reflection of $g_2(u)$ about the $x=0$ axis. The function $g_2(t-u)$ or $g_2(-u+t)$ is $g_2(-u)$ translated to the right by the amount t (t positive) along the u-axis. The function $g_2(t-u)$ is plotted in Figure 2.2c.

The combination of reflection and translation can be further illustrated with an exponential function given by

$$y(x) = \exp(ax)$$

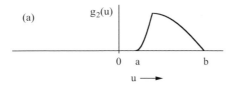

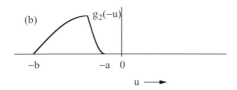

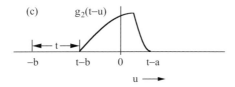

Figure 2.2 The function $g_2(u)$ in (a) is reflected in (b) and translated in (c).

where a is a positive constant. The function $y(x)$ is shown in Figure 2.3a with a $= 1.01$.
Next consider the new function $z(x)$ such that

$$z(x) = y(-x) = \exp(-ax)$$

which is plotted in Figure 2.3b. The function $z(x)$ is $y(x)$ reflected about the vertical
axis through the origin. Consider a third function, $w(x)$, given by

$$w(x) = y(-x + t) = \exp[a(-x + t)]$$

and shown in Figure 2.3c, where it can seen that $w(x)$ is $z(x)$ translated
$t = -0.375$ units, or $y(x)$ reflected about the origin and translated to the left
0.375 units.

The convolution integral as given by Equation 2.1 is the result of reflecting one
of two functions about the vertical axis at the time origin, displacing it a given
distance, multiplying the two functions and integrating the product over the
entire range of the abscissa. As the reflected function is moved to the right or left
for each allowable value of t, the multiplication and integration is repeated.
Figures 2.4 and 2.5 provide two examples that show, step-by-step, how convolution works.

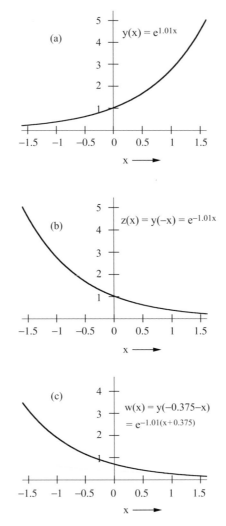

Figure 2.3 Reflecting and translating an exponential function.

2.2.2 A first-order linear system

First-order linear systems have easily understandable properties and can be used to model certain physical systems. In general, physical systems can be thermal, mechanical, electrical or chemical, and each contains some form of resistance or friction. When the system properties are limited to resistance, the behavior of the system can be described by the general first-order linear differential equation with constant coefficients given by

$$a_1 \frac{dy(t)}{dt} + a_0 y(t) = b_0 x(t) \tag{2.7}$$

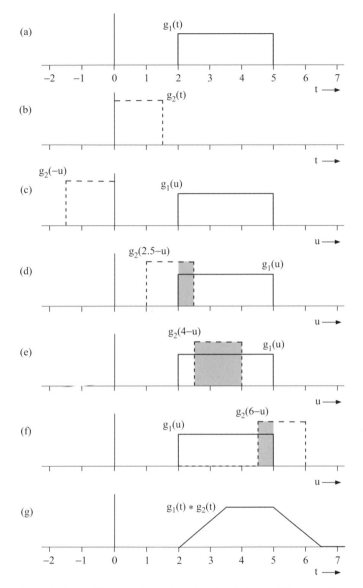

Figure 2.4 Convolution of two rectangular signals. (a) and (b) show the two signals. (c) shows g_2 reflected and g_1 unchanged. (d) shows the reflected g_2 translated to the right 2.5 units, then 4 units in (e) and 6 units in (f). The product of the two functions with time after reflection and translation of g_2 relative to g_1 is shown in (g).

where $x(t)$ represents the input to the system and $y(t)$ the output from the system. It is convenient to divide both sides by a_o and rewrite Equation 2.7 in the form

$$\tau \left(\frac{dy}{dt} \right) + y(t) = x(t) \tag{2.8}$$

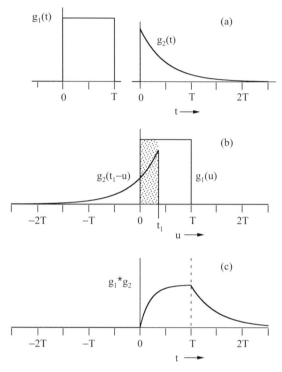

Figure 2.5 Convolution of a rectangular signal with an exponential signal.(a) The two functions. (b) Reflection and translation of the exponential function. (c) The result of the multiplication of the two functions after complete translation.

where $\tau = a_1/a_o$ is a system parameter and $b_o/a_o = 1$ when the input and output have the same scale (as assumed in Equation 2.8). A simple example of a first-order linear system is our previously mentioned liquid-in-glass thermometer in which $y(t)$ is the indicated temperature, $x(t)$ is the environment temperature, and parameter τ is the *time constant*. The glass bulb and liquid in the bulb have mass and, therefore, thermal resistance (that is, it takes time for heat to be conducted into or out of the bulb), as manifested in its time constant. Mathematically, time constant τ is the time required for the thermometer to respond to $1 - (1/e) \approx 0.632$ of a step change in temperature. For example, if the environment temperature suddenly increases $1\,°C$, τ is the time it takes for the thermometer to register a $0.632\,°C$ rise.

The analytic solution to Equation 2.8 can be obtained independently of any prior discussion of convolution. After we have the solution, though, we will recognize that it is in the form of a convolution integral. Multiplying Equation 2.8 by the integrating factor $e^{t/\tau}$ and integrating results in

$$\int_{t=-\infty}^{t=z} d(y\tau e^{t/\tau}) = \int_{t=-\infty}^{t=z} x(t)e^{t/\tau}\,dt$$

which becomes

$$y(z) = \int_{t=-\infty}^{t=z} x(t) \, \frac{e^{-(z-t)/\tau}}{\tau} \, dt.$$

Replacing variables t by u and z by t to match previous notation yields

$$y(t) = \int_{-\infty}^{t} x(u) \, \frac{e^{-(t-u)/\tau}}{\tau} \, du. \tag{2.9}$$

Equation 2.9 is in the form of Equation 2.2 with system function

$$h(t-u) = \frac{e^{-(t-u)/\tau}}{\tau}, \quad u \le t. \tag{2.10}$$

The only difference is in the upper limit, which is now t. The integration extends from all past time up to the present time only. This is because a physical system (the thermometer is an example) "remembers" the past but cannot anticipate the future. More generally, *we cannot know values of the input time series x(u) for any time u later than the current time t.* This maxim is called *physical realizability.* Physical realizability is always connected to the input time series, not the system function. Had we known the system function at the outset, we could have written down Equation 2.9 immediately using the convolution integral and physical realizability without having to formally solve the differential equation.

Let us apply Equation 2.9 to an ordinary thermometer and ask, What is the temperature at time t_1? Figure 2.6 depicts the situation. In general, system function $h(t-u) = e^{-(t-u)/\tau}/\tau$ describes how the measurement of the environment temperature is modified (filtered) by the thermometer to produce the output series of temperature measurements. When adapted to time t_1 it has the form $h(t_1 - u) = e^{-(t_1-u)/\tau}/\tau$, which is the same form as $g_2(t_1 - u)$ in Figure 2.5. Thus Figure 2.6 shows the environment temperature function x(u) and system function $h(t_1 - u)$ from which we conclude that $y(t_1)$, the measured temperature at $t = t_1$, is the exponentially weighted sum of all values of x(u) prior to t_1, and is given mathematically by

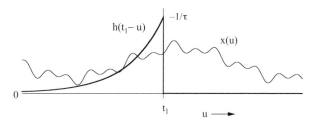

Figure 2.6 Convolution of the weight function h(u) of a liquid-in-glass thermometer with the environment temperature x(u) at time $u = t_1$.

Equation 2.9. Recent values of environment temperature are weighted the most and distant values the least.

2.2.3 More on physical realizability

As noted in Section 2.1, there are two general forms of the convolution integral, Equations 2.1 and 2.2, that can be applied to the analysis of linear systems. Consider the latter first. As discussed above, for physical systems we can perform integration of the input time series x(u) only up to the present time t, beyond which future values of x(u) are unknown. Thus the upper limit of integration is the current time, t, so that for physical systems, in general, Equation 2.2 becomes

$$y(t) = \int_{-\infty}^{t} x(u)\, h(t-u)\, du. \tag{2.11}$$

It can be noted further that only when the value of the argument of the system function is greater than zero is there a contribution to the integration. This is a characteristic of physical systems so that system functions have meaning only when their time argument is greater than or equal to zero.

The limits of integration for Equation 2.1 are somewhat different. Since the argument of the input function is now $(t - u)$, and future values of x$(t - u)$ are not available, it must be that $(t - u) \leq t$; that is, u cannot be negative. Thus, for physical realizability, Equation 2.1 becomes

$$y(t) = \int_{0}^{\infty} h(u)\, x(t-u)\, du. \tag{2.12}$$

Although the argument of the system function is different than in Equation 2.11, the only contribution to integration occurs, again, when its argument is greater than or equal to zero. Equation 2.12 with correct limits can be derived directly from Equation 2.11 by substituting $z = t - u$. Thus both equations are equivalent forms of the convolution integral under the constraint of physical realizability.

2.3 Fourier transforms for analog data

In Chapter 1 our primary interest was in analyzing digital signals of finite record length. One of the more important results was the digital Fourier transform pair given by Equations 1.61 and 1.62 that was derived in Section 1.5.5. In this section we will derive two new Fourier transform pairs, the first for a finite analog record and the second for an infinite analog record. The starting point for both transforms is the equation for the complex amplitude spectrum.

We first insert Δt into the numerator and denominator of the summation coefficient and the exponential term of Equation 1.61 to obtain

$$S'_m = \frac{1}{N\Delta t} \sum_{n=-[N/2]}^{n=[(N-1)/2]} x_n \Delta t \, e^{-i2\pi mn\Delta t/N\Delta t}, \quad m = -[N/2], \ldots, 0, \ldots [(N-1)/2]$$

$$(1.61')$$

where S'_m is the complex amplitude of the m-th harmonic and the prime indicates the implied Fourier coefficients are one-half the values defined in Table 1.1, except that $A'_0 = A_0$ and $A'_{N/2} = A_{N/2}$ (N even). In addition, for convenience, the limits of the summation have been changed so that x_n is centered about $n = 0$ for N odd and displaced one time unit for N even where $[q]$ means truncation of q.

Next, let Δt tend to zero and N tend to infinity in such a way that $N\Delta t = T$, where T is the finite temporal length of the record. Simultaneously, let $n\Delta t$ tend to a point in time denoted by t, $x_n \Delta t$ tend to $x(t)dt$, and the summation tend to an integral. The result of applying these limits to Equation 1.61' is

$$S'_m = \frac{1}{T} \int_{-T/2}^{T/2} x(t) \, e^{-i2\pi mt/T} \, dt, \quad -\infty \le m \le \infty. \qquad (2.13)$$

Correspondingly, Equation 1.62 tends to

$$x(t) = \sum_{m=-\infty}^{\infty} S'_m \, e^{i2\pi mt/T}, \quad -T/2 \le t \le T/2 \qquad (2.14)$$

thereby completing the Fourier transform pair for a finite analog record. Equation 2.14 is the inverse Fourier transform of Equation 2.13. We can think of $x(t)$ as an original analog signal that could have been digitally sampled to produce x_n.

Extension of Equations 2.13 and 2.14 to an infinite record length can be approached in the following way. The difference in frequency between adjacent harmonic frequencies is given by

$$\Delta f = \frac{m+1}{T} - \frac{m}{T} = \frac{1}{T}.$$

As T tends to infinity, Δf tends to df, and the harmonic frequencies m/T tend to a continuous variation in frequency denoted by the variable f. By dividing Equation 2.13 by $1/T$, the left side becomes $S'_m/(1/T)$, which accounts for the amplitude variation in the frequency width $1/T$, that is, amplitude per unit bandwidth or *amplitude density*, which, as T tends to infinity, is written $X(f)$, as in Equation 2.15 below. The concept of amplitude density may be difficult to grasp. That we are forced to deal with amplitude density is a consequence of having finite amplitudes becoming increasingly closely spaced along the frequency axis as T increases without bound.

However, having finite amplitudes at frequencies with infinitesimal separation losses mathematical meaning. The solution is to have a continuum of amplitudes resulting in an amplitude density spectrum in the same way that a continuum of harmonic variances resulted in a variance density spectrum, briefly discussed in Section 1.1 (and that will be developed further in Chapter 5). Amplitude density is analogous to probability density. We cannot realize a value of probability at a particular value of the independent variable, but we can realize a value of probability if we integrate the probability density function over a range of the independent variable. Similarly, we cannot realize a value of amplitude at a particular frequency, but we can realize a value of amplitude if we integrate the amplitude density function over a range in frequency.

To provide a consistent transform pair we divide S'_m in Equation 2.14 by $1/T$ and multiply the exponential term by $1/T$, and, again, let T tend to infinity. The result of these limiting operations is the Fourier transform pair for an infinite analog record:

$$X(f) = \int_{-\infty}^{\infty} x(t)\, e^{-i2\pi ft}\, dt, \qquad -\infty \leq f \leq \infty \qquad (2.15)$$

and

$$x(t) = \int_{-\infty}^{\infty} X(f)\, e^{i2\pi ft}\, df. \qquad -\infty \leq t \leq \infty. \qquad (2.16)$$

Equation 2.16 is the inverse Fourier transform of Equation 2.15.

Comparable derivations of Equations 2.13 through 2.16 can be found in Koopmans (1974, pp. 19–21, 23–25) and Jenkins and Watts (1968, pp. 23–24), among other sources. In order that these Fourier integrals exist it is assumed that $\int_{-\infty}^{\infty} |x(t)|dt < \infty$ and $\int_{-\infty}^{\infty} |X(f)|df < \infty$ (Koopmans, 1974, p. 24). Equations (2.15) and (2.16) are commonly used in theoretical investigations, an example of which is given in the next section.

Equivalent expressions for Equations 2.15 and 2.16 using angular frequency instead of ordinary frequency can be written down directly by substituting $\omega = 2\pi f$. Thus,

$$X(\omega) = \int_{-\infty}^{\infty} x(t)\, e^{-i\omega t}\, dt, \qquad -\infty \leq \omega \leq \infty \qquad (2.17)$$

and

$$x(t) = \frac{1}{2\pi} \int_{-\infty}^{\infty} X(\omega)\, e^{i\omega t}\, d\omega. \qquad -\infty \leq t \leq \infty. \qquad (2.18)$$

Of course, either amplitude density function Equation 2.15 or 2.17 can be used and the relation between them is $X(f) = 2\pi\, X(\omega)$.

2.4 The delta function

In this section we derive one example of a class of functions called *generalized functions*. In dealing with generalized functions, one is not so much concerned with their behavior on a point-by-point basis but rather with their effect on the values of integrals and other functionals in which they appear. All generalized functions have derivatives of all orders and each has a Fourier transform (Lumley, 1972, p. 159). One of these generalized functions is the *Dirac delta function* or simply *delta function* [after P.A.M. Dirac, renowned English atomic physicist; see Dirac (1947, pp. 58–62) for additional details on the development of this function]. The delta function is very useful in the mathematical analysis of physical systems.

We begin the derivation by obtaining the Fourier transform of a rectangular function of amplitude as shown in Figure 2.7a. From Equation 2.15 we have

$$X(f) = a \int_{-b}^{b} e^{-i2\pi ft} \, dt \tag{2.19a}$$

which, because of even symmetry about $t = 0$, reduces to

$$X(f) = a \int_{-b}^{b} \cos(2\pi ft) \, dt$$
$$= 2ab \frac{\sin(2\pi fb)}{2\pi fb}. \tag{2.19b}$$

From this equation we see that the amplitude density function $X(f) = 0$ at $f = k/(2b)$, $k = \pm 1, \pm 2, \ldots$, while at $f = 0$, $X(0) = 2ab$ following application of l'Hopital's rule. Equation 2.19b, plotted in Figure 2.7b, is the familiar diffraction function, a bi-directional damped sinusoid.

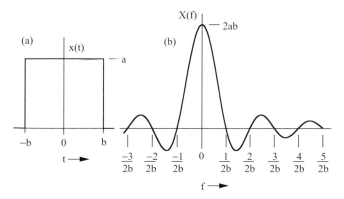

Figure 2.7 A rectangular function (a) and its Fourier transform (b).

Consider the case where x(t) has unit height, that is, a = 1. Then Equation 2.19b becomes

$$X(f) = 2b \frac{\sin(2\pi fb)}{2\pi fb}. \tag{2.20}$$

Similar to Figure 2.7, the rectangular function and its Fourier transform are shown in Figure 2.8a. Now increase the width of the rectangular function by, say, a factor of three and compute its Fourier transform. The results are shown in Figure 2.8b, in which the increased width of the time function results in reduced separation between consecutive zero crossings. In the limit as b → ∞, that is, the time function extends indefinitely in either direction, the spacing between consecutive zero crossings approaches zero so that X(f) = 0 everywhere except at f = 0, where X(0) → ∞. Figure 2.8c shows the result of this process: a time function of unit height and infinite extent and a frequency function of infinitesimal thickness and infinite height, the latter function represented by the vertical line and arrow. This is a common definition of the delta function and has the special notation δ(f).

The above result is not too surprising. As b increases without bound in Figure 2.8, x(t) approaches a constant value, the frequency of which is zero. Simultaneously, the area under x(t) increases without bound so that, in the limit, the amplitude density is concentrated at f = 0 and X(f = 0) is infinite. From Figure 2.8c we conclude that, mathematically,

$$X(f) = \int_{-\infty}^{\infty} 1 \cdot e^{-i2\pi ft} \, dt = \delta(f). \tag{2.21}$$

We can anticipate that the Fourier transform of Equation 2.21 is

$$x(t) = \int_{-\infty}^{\infty} \delta(f) \, e^{i2\pi ft} \, df = 1. \tag{2.22}$$

Equations 2.21 and 2.22 comprise the Fourier transform pair between a function of unit value in the time domain and a delta function in the frequency domain.

To determine the area under the diffraction function we integrate X(f) in Equation 2.20 from −∞ to ∞ and make use of the known integral

$$\int_{-\infty}^{\infty} \frac{\sin z}{z} \, dz = \pi$$

to show that the result is unity and independent of b. Since δ(f) is formed in the limit as b tends to ∞, the area under δ(f) is also unity. Thus,

$$\int_{-\infty}^{\infty} X(f) df = \int_{-\infty}^{\infty} 2b \frac{\sin(2\pi fb)}{2\pi fb} \, df = 1 \tag{2.23a}$$

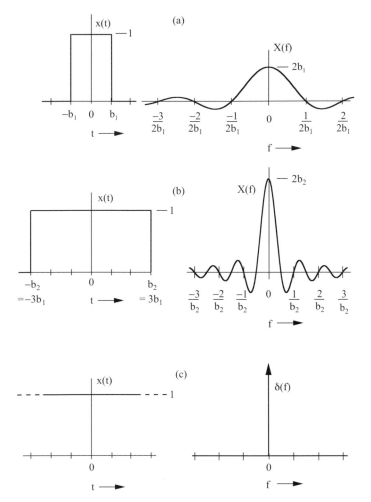

Figure 2.8 Evolution of x(t) toward a constant (unit value) simultaneously with X(f) toward a delta function $\delta(f)$.

and

$$\int_{-\infty}^{\infty} \delta(f)df = 1. \tag{2.23b}$$

Next, consider the case in which the area under x(t) in Figure 2.7 is one, that is, $2ab = 1$. If b $\rightarrow$ 0 in such a way that the area remains at unit value, a must increase without bound. This evolution can be followed in Figure 2.9. Panel (a) shows a rectangular function with unit area and its Fourier transform with unit amplitude density at $f = 0$. Panel (b) shows a narrower and taller time function with unit area and associated amplitude density function but with much wider separation between

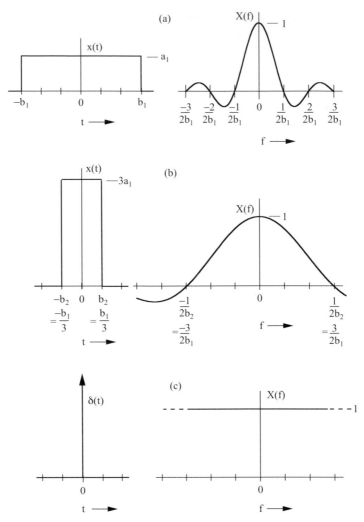

Figure 2.9 Evolution of $x(t)$ toward a delta function $\delta(t)$ simultaneously with $X(f)$ toward a constant (unit value). The area under $x(t) = 1$, i.e., $2a_1b_1 = 1$.

zero crossings and, again, unit amplitude density at $f = 0$. In the limiting case in panel (c), the time function has infinitesimal width and infinite height such that unit area is preserved while the amplitude density function has infinite separation between zero crossings and unit value everywhere. We conclude from the process in Figure 2.9 that

$$X(f) = \int_{-\infty}^{\infty} \delta(t)\, e^{-i2\pi ft}\, dt = 1. \tag{2.24}$$

Thus it takes an amplitude density constant with frequency to reconstruct a delta function in the time domain. The Fourier transform of Equation 2.24 is

$$x(t) = \int_{-\infty}^{\infty} 1 \cdot e^{i2\pi ft} \, df = \delta(t) \tag{2.25}$$

so that Equations 2.24 and 2.25 provide the Fourier transform pair between a delta function in the time domain and a constant in the frequency domain. An important implication of Equation 2.24 is that if a data set contains a very large glitch or spike [$x(t)$ tends to $\delta(t)$], the periodogram will contain approximately uniform variance (called noise) across all frequencies due to the spike.

In terms of angular frequency, ω, expressions parallel to Equations 2.21, 2.22, 2.24, and 2.25 are:

$$X(\omega) = \int_{-\infty}^{\infty} 1 \cdot e^{-i\omega t} \, dt = 2\pi \, \delta(\omega) \tag{2.26}$$

$$x(t) = \frac{1}{2\pi} \int_{-\infty}^{\infty} 2\pi \, \delta(\omega) \, e^{i\omega t} \, d\omega = 1 \tag{2.27}$$

$$X(\omega) = \int_{-\infty}^{\infty} \delta(t) \, e^{-i\omega t} \, dt = 1 \tag{2.28}$$

and

$$x(t) = \int_{-\infty}^{\infty} 1 \cdot e^{i\omega t} \, d\omega = \delta(t). \tag{2.29}$$

These equations can be derived in a manner similar to those involving frequency by applying the Fourier transform of Equation 2.17 to the rectangular function in Figure 2.7. The comparable zero crossings will occur at integer multiples of $2\pi/(2b) = \pi/b$ and the amplitude will be $2ab/(2\pi) = ab/\pi$. Increasing the width of the rectangle indefinitely while keeping its height at unit value and taking its Fourier transform, as in Figure 2.8, yields Equations 2.26 and 2.27; fixing the area of the rectangle at unit value as its width approaches zero and taking its transform, as in Figure 2.9, yields Equations 2.28 and 2.29.

Three properties that are a consequence of the definition of the delta function are:

$$(1) \quad \delta(t) = \begin{cases} 0, & t \neq 0 \\ \infty, & t = 0 \end{cases} \quad (\text{dimension is time}^{-1}) \tag{2.30}$$

$$(2) \quad \int_{-\infty}^{\infty} \delta(t) dt = 1 \tag{2.31}$$

and

$$(3) \quad \int_{-\infty}^{\infty} g(t)\, \delta(t_0 - t)\, dt = g(t_0). \qquad (2.32)$$

Property (2) is a direct consequence of applying property (1) to Equation 2.24 and recalling that the area under the rectangle in Figure 2.9 was unity. The third property is called the "sifting" or "sampling" property and is simply a convolution integral. It can be used to sample the value of a function, here $g(t)$, at time t_0. Note that property (3) includes properties (1) and (2). There are a number of other properties (Dirac, 1947; Cooper and McGillem, 1967) but only those above are used in this chapter. Because the argument of a delta function is arbitrary, the properties above apply as well to delta functions in the frequency domains (f and ω). In the f-domain the dimension of the delta function is inverse ordinary frequency and in the ω-domain it is inverse angular frequency.

In application to a finite record of digital data, the following can be observed through simulation. A single finite spike or glitch located anywhere in a data sequence that is otherwise constant in value results in a periodogram of essentially constant variance, as mentioned above and in agreement with Equation 2.24. If there are two finite spikes in the data set, the periodogram is no longer flat but contains a pattern of small and large variances that is dependent on phase angle relationships of the two sinusoids at each harmonic in accord with Equation 1.48. As the number of spikes increases, the periodogram becomes more chaotic. Thus one cannot remove a portion of the observed variance at each harmonic in order to rid the spectrum of the effect of the spikes in the data. The solution is to remove the spikes directly through the use of some type of nonlinear filter, for example, the median filter investigated by Brock (1986).

2.5 Special input functions

One way to determine the properties of a linear system in the time domain is to input one of two special functions called the *impulse function* and the *step function*. In an experimental set up it is usually more practical to introduce a step change in input to a measurement system than an impulse. Here we consider the mathematical aspects of these functions as inputs to a system and the resulting outputs.

2.5.1 Impulse function

Let the input $x(t) = k\delta(t)$ be defined as an impulse function in which k serves to scale the unit area that results from integration of a delta function (Equation 2.31) to other than unit value, that is, to value k. If k is unity, then $k\delta(t) = 1\cdot\delta(t)$ is the *unit impulse function* and is equivalent to the delta function.

Now let $\delta(t)$ be the input to a linear system. From Equation 2.1 the output is

$$y(t) = \int_{-\infty}^{\infty} h\,(u)\,\delta(t-u)\,du = h\,(t) \qquad (2.33)$$

making use of the sampling property (Equation 2.32) of a delta function. Thus $y(t)$ is the response to a unit impulse. Because $h(t)$, referred to earlier as the system function, is identical to $y(t)$, it is called the *impulse response function*; that is, $h(t)$ is the response to introducing an impulse into the system. From a theoretical viewpoint a unit impulse could be introduced into any linear physical system to yield its impulse response function without knowing the integro-differential equation that describes the system. It is usually desirable, though, to know the mathematical form of $h(t)$. While it is possible, in principle, to model $h(t)$ given the numerical output from the unit impulse function, it is better to know, in advance, the integro-differential equation for the system. Then the form of $h(t)$ can be determined directly and only the system parameter values (coefficients) need to be estimated.

We apply this idea to the first-order linear differential equation given by Equation 2.8, the solution of which is Equation 2.9. If $x(u)$ in Equation 2.9 is replaced by $\delta(u)$ then, from Equation 2.33,

$$y(t) = h(t) = (1/\tau)\,e^{-t/\tau}, \quad t \geq 0 \qquad (2.34)$$

as shown in Figure 2.10. To determine τ, we find $h(\tau) = (1/\tau)\,e^{-1} = 0.368(1/\tau)$ on the vertical axis of Figure 2.10b, move to the right until it intersects the curve, then drop to the horizontal time axis where $t = \tau$, the time constant. Alternatively, τ can be determined directly from $y(0) = h(0) = 1/\tau$, but this is less useful experimentally. The use of the unit impulse function in determining τ is not possible for physical systems because of the requirement for a pulse of infinite amplitude. The unit impulse input approach can work in certain electronic systems in which the pulse is of sufficiently large magnitude.

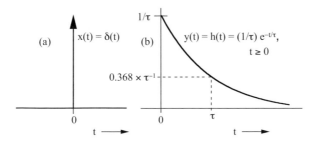

Figure 2.10   (a) Impulse input to a first-order linear system. (b) Output from system.

2.5.2 Step function

A more satisfactory approach to estimate parameter values of a system function from an application viewpoint is to use a step function. Although it also demands an instantaneous change, the change is finite.

The *unit step function* is defined as

$$U(t) = \begin{cases} 0, & t < 0 \\ 1, & t \geq 0. \end{cases} \tag{2.35}$$

Let $x(t) = qU(t)$ where q is a constant with dimensions of $x(t)$ and which serves to scale the unit step function to the desired level. If $x(t)$ is the input to the system, the output is then

$$y(t) = q \int_0^t U(u)\, h(t - u)\, du \tag{2.36}$$

where the lower limit is a consequence of $U(u) = 0$ for $u < 0$ from Equation 2.35 and the upper limit is a consequence of physical realizability. Following a change in variable, Equation 2.36 can be equivalently written

$$y(t) = q \int_0^t h(u)\, U(t - u)\, du. \tag{2.37}$$

Substituting Equation 2.35 into Equation 2.37 yields

$$y(t) = q \int_0^t h(u)\, du. \tag{2.38}$$

To find $y(t)$ we differentiate the output so that

$$\frac{dy}{dt} = qh(t). \tag{2.39}$$

Experimentally, $y(t)$ would have to be modeled to allow mathematical differentiation.

In the case of a first-order linear system, $h(t)$ is given by Equation 2.34 so that we can write the output $y(t)$ in Equation 2.38 as

$$y(t) = q\left(1 - e^{-t/\tau}\right), \quad t \geq 0. \tag{2.40}$$

Equations 2.35 and 2.40 are plotted in Figure 2.11.

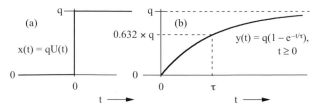

Figure 2.11 (a) Step input to first-order linear system. (b) Output from system.

Now consider a numerical example in which a step function is used to determine the time constant of a liquid-in-glass thermometer. Let the thermometer initially be at room temperature $= 23\,°C$. It is then instantly immersed in a water bath at $43\,°C$ with attentive eyes reading the temperature scale to produce the solid curve shown in Figure 2.12. We need to take into account that the reference value is nonzero and that $U(t)$ represents the departure from this value. Therefore, from Equation 2.40, the output can be written

$$y(t) = q\left(1 - e^{-t/\tau}\right) + T_o \qquad (2.41)$$

or, in $°C$,

$$y(t) - 23 = 20\left(1 - e^{-t/\tau}\right). \qquad (2.42)$$

From Equation 2.42 we see that the time constant τ is the time required for the difference between the thermometer temperature and the water bath temperature to reach $(1 - e^{-1}) = 0.632$ of the initial difference. That is,

$$y(t = \tau) = 20(1 - e^{-1}) + 23$$
$$= 12.64 + 23$$
$$y(t = \tau) = 35.64°C.$$

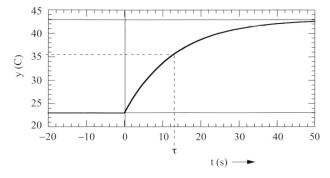

Figure 2.12 Response of a thermometer to a 20 degree Celsius step change in temperature.

If one draws a horizontal line from $35.64\,°C$ until it intersects the temperature curve and from there a vertical line downward to the time axis, the latter intersection corresponds to the time constant, that is, $\tau = 13\,s$.

2.6 The frequency response function

The Fourier transform of the input of a linear system yields the amplitude density spectrum of the input as given by Equation 2.15, and similarly for the output. The Fourier transform of the system function $h(t)$ shows how the Fourier amplitudes and phase angles of input signal $x(t)$ are modified by the system to produce the Fourier amplitudes and phase angles in the output signal $y(t)$. Accordingly, the Fourier transform of system function $h(t)$ is given by

$$H(f) = \int_{-\infty}^{\infty} h(t)\,\exp(-i2\pi ft)\,dt \tag{2.43}$$

where $H(f)$ is called the *frequency response function* and is, in general, complex and contains information about changes in amplitude and phase angle between the input and output of a linear system. The above integral applies to mathematical systems; in the case of physical systems, the lower limit of integration is zero due to physical realizability. Equation 2.43 can be written also as

$$H(f) = G(f)\,e^{i\phi(f)} = G(f)[\cos\phi(f) + i\,\sin\phi(f)] \tag{2.44}$$

where $G(f)$ is called the *gain function* or *gain factor* or simply *gain* and $\phi(f)$ is called the *phase function* or *phase shift*. We can think of $H(f)$ as a vector in the complex plane as shown in Figure 2.13, where Im is the imaginary axis and Re is the real axis. $G(f)$ is the modulus or absolute value of $H(f)$, which can be determined by multiplying $H(f)$ by its complex conjugate and taking its square root. Thus,

$$G(f) = |H(f)| = [H(f)\,H*(f)]^{1/2}. \tag{2.45}$$

To understand the meanings of gain and phase shift, consider an input sinusoid given by

$$x(t) = a\,\cos(2\pi ft). \tag{2.46}$$

Let us pass this signal through a first-order linear system, the output of which is given by

$$y(t) = \int_{-\infty}^{t} x(u)\,\frac{e^{-(t-u)/\tau}}{\tau}\,du \tag{2.9}$$

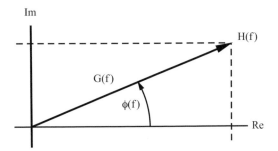

Figure 2.13   The gain G(f) and phase $\phi(f)$ components of the frequency response function H(f) in the complex plane.

as derived in Section 2.2.2. Inserting the input sinusoid into Equation 2.9 and integrating yields

$$y(t) = \frac{a}{[1 + (2\pi f \tau)^2]} \left[\cos(2\pi ft) + (2\pi f \tau) \sin(2\pi ft) \right]$$

$$= \frac{a}{[1 + (2\pi f \tau)^2]^{1/2}} \cos\left[2\pi ft - \tan^{-1}(2\pi f \tau) \right] \qquad (2.47)$$

$$= a\, G(f) \cos[2\pi ft + \phi(f)]$$

where

$$G(f) = \left[1 + (2\pi f \tau)^2 \right]^{-1/2}$$

and

$$\phi(f) = -\tan^{-1}(2\pi f \tau).$$

From Equation 2.47 it is evident that the gain G(f) is the ratio of the amplitude of the output sinusoid to the amplitude of the input sinusoid at frequency f, and the phase shift $\phi(f)$ is the angle by which the phase angle of the input signal (here 0) is shifted in the output signal at frequency f. Figure 2.14 shows the input and output sinusoids in Equations 2.46 and 2.47.

2.6.1 First-order linear system

In this section we will derive H(f), G(f) and $\phi(f)$ for the first-order linear system in Equation 2.8. The impulse response function is already known. Therefore, from

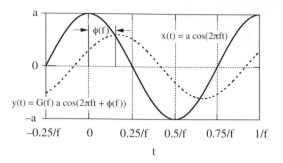

Figure 2.14   An example that illustrates the meaning of gain G(f) and phase angle φ(f). The solid curve is input x(t) and the dashed curve is output y(t).

Equations 2.43 and 2.34, the frequency response function is

$$H(f) = \int_0^\infty h(t)\, e^{-i2\pi ft}\, dt$$

$$= (1/\tau) \int_0^\infty e^{-t/\tau}\, e^{-i2\pi ft}\, dt \qquad (2.48)$$

$$H(f) = (1 + i2\pi f\tau)^{-1}.$$

From Equation 2.45,

$$G(f) = [H(f)\, \overset{*}{H}(f)]^{1/2}$$

$$= \left[\frac{1}{1 + i2\pi f\tau} \times \frac{1}{1 - i2\pi f\tau} \right]^{1/2} \qquad (2.49)$$

$$G(f) = \left[1 + (2\pi f\tau)^2 \right]^{-1/2}.$$

From Equation 2.44,

$$e^{i\phi(f)} = \frac{H(f)}{G(f)}$$

so that

$$\cos\phi(f) + i\,\sin\phi(f) = \frac{\left[1 + (2\pi f\tau)^2 \right]^{1/2}}{1 + (i2\pi f\tau)}.$$

After equating real and imaginary parts on the left to those on the right, we obtain

$$\phi(f) = -\tan^{-1}(2\pi f\tau). \qquad (2.50)$$

As expected, the formulas for the gain and phase functions derived here are identical to those we derived in the previous section by way of an example. A summary of the response characteristics of a first-order linear system is given in Table 2.1.

Table 2.1   Responses for a first-order linear system.

Unit impulse response	Unit step response	Frequency response	Gain function	Phase function
$y(t) = h(t) =$ $(1/\tau)e^{-t/\tau},$ $(t \geq 0)$	$y(t) =$ $1 - e^{-t/\tau},$ $(t \geq 0)$	$H(f) =$ $(1 + i2\pi f\tau)^{-1}$	$G(f) =$ $[1 + (2\pi f\tau)^2]^{-1/2}$	$\phi(f) =$ $-\tan^{-1}(2\pi f\tau)$

The gain and phase functions are plotted in Figures 2.15 and 2.16, respectively. It should be noted that the phase shift cannot be less than $-90°$ and approaches this value asymptotically as frequency increases. When the period of the input sinusoid is equal to 2π times the time constant $(1/f = 2\pi\tau)$ the phase angle is $-45°$ and the gain is 0.707 (i.e., $1/\sqrt{2}$). Also, when the input period is equal to the time constant $(1/f = \tau)$ the phase angle is $-81°$, almost its maximum negative value. Simultaneously, the gain is reduced to 0.157.

As a numerical example, let us say that we are observing ocean temperature at a depth of two meters and we want to know how a thermometer will respond to a sinusoidal water temperature oscillation with a one-second period if the thermometer's time constant is 0.5 s. From Table 2.1 $G(f = 1 \, \text{s}^{-1}) = [1 + (2\pi f\tau)^2]^{-1/2} = [1 + (2\pi 0.5)^2]^{-1/2} = 0.303$ and $\phi(f = 1 \, \text{s}^{-1}) = -72.34°$. The actual and observed

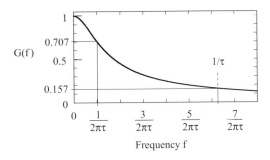

Figure 2.15   Gain function G(f) for a first-order linear system.

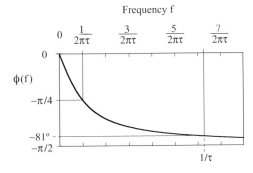

Figure 2.16   Phase function $\phi(f)$ for a first-order linear system

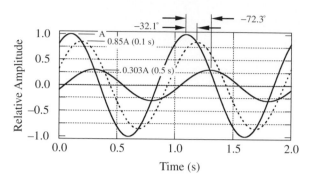

Figure 2.17 Example of a sinusoidal input with period 1 s and amplitude A and resulting outputs from a first-order system for two different time constants τ given in parentheses.

temperature variations are shown in Figure 2.17. As is evident from this figure, the response to a one cycle per second sinusoid is quite small; the output amplitude is only 30% of the input amplitude and there is a large lag relative to the period of the sinusoid.

Alternatively, consider the case when the time constant is 0.1 s. Then $G(f = 1\,s^{-1}) = 0.850$ and $\phi(f = 1\,s^{-1}) = -32.14°$. Figure 2.17 shows the response is much more favorable. One might consider as a rule-of-thumb that to obtain a good representation of the signal at a desired frequency for a first-order linear system, the time constant should be at most one-tenth the period of the oscillation.

2.6.2 Integration

Another example of a linear system is an integrator. This could be an electronic or mechanical integrating device or mathematical integration. For an integrating system, the output has the simple form

$$y(t) = \int_{-\infty}^{t} x(u)\, du. \tag{2.51}$$

To find the impulse response function for an integrator, we substitute the unit impulse function $\delta(u)$ for $x(u)$ in Equation 2.51 and obtain

$$h(t) = \begin{cases} 1, & t \geq 0 \\ 0, & t < 0 \end{cases} \tag{2.52}$$

which is identical to the unit step function in Equation 2.35.

The frequency response function is

$$H(f) = \int_{0}^{\infty} h(t) \exp(-i2\pi ft)\, dt$$

$$H(f) = (i2\pi f)^{-1} \tag{2.53}$$

from which the gain is

$$G(f) = [H(f) \, H*(f)]^{1/2} = (2\pi f)^{-1}. \qquad (2.54)$$

From Equation 2.44, the phase angle is

$$\phi(f) = -\pi/2. \qquad (2.55)$$

Regardless of the frequency, there is always a phase shift of 90° between the input and output (because one is a sine and the other a cosine). The frequency response function, gain, and phase angle are undefined for zero frequency. This is because Equation 2.53 is infinite, reflecting continuous integration of a constant input, which, if not zero, would result in an infinite value as t tends to infinity.

Now consider a particular case of a rain gauge that measures accumulated rain with time. The input signal x(t) to the rain gauge is rain intensity or rain rate in mm/h. Let

$$x(t) = a \begin{cases} 1 - \cos(2\pi t/T), & 0 \le t \le 3T \\ 0, & \text{elsewhere} \end{cases} \qquad (2.56)$$

with T = 1 hour. The dashed line in Figure 2.18 is the rain rate with a = 25 mm/h. Thus Equation 2.56 represents a rain event occurring over a three-hour period, in which the peak intensity of 25 mm/h repeats itself at the midpoint of each hour and the rain rate decreases to zero on the hour in a sinusoidal manner. Of course, no real meteorological event would yield an analytical rain rate such as Equation 2.56; it is employed only to illustrate system integration. From the

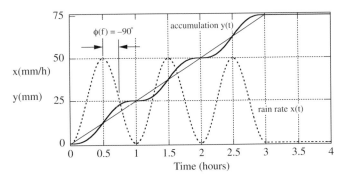

Figure 2.18 Example of input to and output from an integrating device, in this case, a weighing rain gauge. The input rain rate x(t) is accumulated according to y(t). φ(f) is phase angle.

convolution formula Equation 2.1, the system output, or accumulated rain, at time t is

$$y(t) = \int_{-\infty}^{\infty} x(u)\, h(t-u)\, du$$

$$= \int_0^t x(u)\, du$$

$$= a \int_0^t [1 - \cos(2\pi u/T)]\, du$$

$$= \begin{cases} a \int_0^t [1 - \cos(2\pi u/T)]\, du, & 0 \le t \le 3T \\[2ex] a \int_0^{3T} [1 - \cos(2\pi u/T)]\, du + a \int_{3T}^t 0\, du, & t > 3T \end{cases}$$

$$y(t) = a \begin{cases} t - (T/2\pi)\sin(2\pi t/T), & 0 \le t \le 3T \\ 3T, & t > 3T \end{cases} \tag{2.57}$$

The same result would be obtained with the first integrand being h(u) x(t − u).

The heavy solid line in Figure 2.18 is the accumulation y(t) with, again, a = 25 mm/h. To verify the gain and phase relationships, we first subtract the nonfluctuating components. Thus, we subtract a from x(t) and at from y(t) over the range $0 \le t \le 3T$ and find that the ratio of the amplitude of the output sinusoid to the amplitude of the input sinusoid, that is, the gain, is $T/2\pi$ as predicted by Equation 2.54 and [y(t) − at] lags [x(t) − a] by 90° (i.e., $\phi(f) = -\pi/2$) as predicted by Equation 2.55. The phase angle is shown in Figure 2.18 along with light solid line at.

2.7 Fourier transform of the convolution integral

Recall from Section 2.1 that the solution to a linear integro-differential equation is given by the convolution integral

$$y(t) = \int_{-\infty}^{\infty} x(u)\, h(t-u)\, du. \tag{2.1}$$

Consider now the Fourier transform (denoted by FT) of Equation 2.1. From Equation 2.15,

$$FT[y(t)] = Y(f) = \int_{-\infty}^{\infty} \left[\int_{-\infty}^{\infty} x(u)\, h(t-u)\, du \right] e^{-i2\pi ft}\, dt. \tag{2.58}$$

After interchanging the order of integration we have

$$Y(f) = \int_{-\infty}^{\infty} x(u)\, e^{-i2\pi fu} \left[\int_{-\infty}^{\infty} h(t-u)\, e^{-i2\pi f(t-u)}\, d(t-u)\right] du$$

$$= \int_{-\infty}^{\infty} x(u)\, e^{-i2\pi fu}\, H(f)\, du \qquad (2.59)$$

$$Y(f) = X(f)\, H(f)$$

or

$$H(f) = \frac{Y(f)}{X(f)}. \qquad (2.60)$$

That is, the ratio of the output amplitude spectrum to the input amplitude spectrum at frequency f is the frequency response function.

Convolution in the frequency domain parallels convolution in the time domain and is given by

$$Z(f) = \int_{-\infty}^{\infty} Q(v)\, W(f-v)\, dv \qquad (2.61)$$

where f and v are ordinary frequency variables. If we apply Equation 2.58, except we take the inverse Fourier transform and integrate with respect to frequency, we have

$$FT^{-1}\,[Z(f)] = z(t) = \int_{-\infty}^{\infty} \left[\int_{-\infty}^{\infty} Q(v)\, W(f-v)\, dv\right] e^{i2\pi ft}\, df. \qquad (2.62)$$

The sign of the exponent in the inverse Fourier transform is positive. After interchanging the order of integration and following Equation 2.59, except in the frequency domain, Equation 2.62 reduces to

$$z(t) = q(t)\, w(t) \qquad (2.63)$$

where $q(t)$ and $w(t)$ are the inverse Fourier transforms of $Q(f)$ and $W(f)$, respectively.

Equations 2.1 and 2.59 tell us that the Fourier transform of the convolution integral in the time domain is equivalent to the product of the Fourier transforms of the individual terms in the frequency domain. Equations 2.61 and 2.63 tell us the opposite also is true: the inverse Fourier transform of the convolution integral in the frequency domain is equivalent to the product of the inverse Fourier transforms of the individual terms in the time domain. Expressed mathematically, and using now the notation in Equation 2.6, we have the general relations

$$FT\,[g_1(t)*g_2(t)] = FT[g_1(t)]\, FT\,[g_2(t)] = G_1(f)\, G_2(f) \qquad (2.64)$$

or

$$FT^{-1}[G_1(f)\,G_2(f)] = g_1(t) * g_2(t) \tag{2.65}$$

and

$$FT^{-1}[G_1(f) * G_2(f)] = FT^{-1}[G_1(f)]\,FT^{-1}[\,G_2(f)] = g_1(t)\,g_2(t) \tag{2.66}$$

or

$$FT\,[g_1(t)\,g_2(t)] = G_1(f) * G_2(f). \tag{2.67}$$

These results show that Fourier transformation converts convolution (multiplication) in one domain to multiplication (convolution) in the other domain.

If physical realizability is involved, the upper limit of integration in Equation 2.1 is t and Equation 2.59 becomes

$$Y(f) = X(f,t)\,H(f) \tag{2.68}$$

where

$$X(f,t) = \int_{-\infty}^{t} x(u)\,e^{-i2\pi fu}\,du. \tag{2.69}$$

The above derivation has been done between the time and ordinary frequency domains. If the convolution integral in Equation 2.1 had been transformed to the ω-domain, the formula equivalent to Equation 2.59 would be

$$Y(\omega) = X(\omega)\,H(\omega). \tag{2.70}$$

Replacing the ordinary frequencies f and v in Equation 2.61 by angular frequency variables provides the formula for convolution in the angular frequency domain. Taking its inverse Fourier transform using Equation 2.18 results in Equation 2.63 but with the multiplicative coefficient 2π on the right-hand side. The derivation is an interesting and useful exercise.

2.8 Linear systems in series

So far we've dealt with an input signal passing through a linear physical system resulting in an output signal as described in Figure 2.1. In this section we consider the possibility that the input signal passes through more than one physical system, in fact, any number of systems. As long as the properties of each system are known, we can find the output signal given the input signal and vice versa providing that overall system noise is negligible. The signal flow for a sequence of n systems is shown in Figure 2.19, where the first input $x(t) = x_1(t)$ passes through the system with impulse response or system function h_1, its output $y_1(t)$ becomes the input to the next system with impulse response h_2 and so on, until the final output $y(t) = y_n(t)$ is realized.

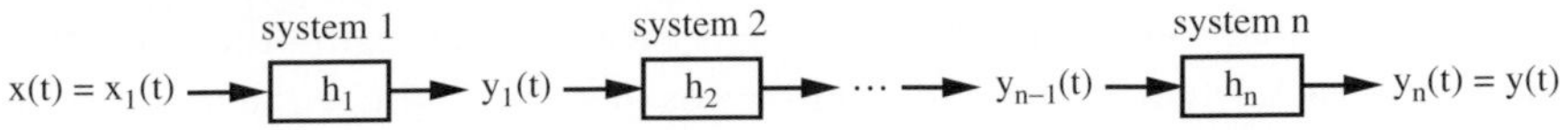

system 1   system 2   system n

$x(t) = x_1(t)$ →[h_1] → $y_1(t)$ →[h_2] → $\cdots$ → $y_{n-1}(t)$ →[h_n] → $y_n(t) = y(t)$

Figure 2.19   The signal flow for a sequence of n systems.

From Equation 2.58, for two systems in series

$$\text{system 1}\quad Y_1(f) = H_1(f) \cdot X(f)$$

$$\text{system 2}\quad Y_2(f) = H_2(f) \cdot Y_1(f)$$

or

$$\text{system 2}\quad Y_2(f) = H_1(f) \cdot H_2(f) \cdot X(f).$$

Repeated application of Equation 2.58 will show that for n linear systems in series

$$Y(f) = H_1(f) \cdot H_2(f) \ldots H_n(f) \cdot X(f) \tag{2.71}$$

in which the overall frequency response function is the product of the individual frequency response functions, that is,

$$H(f) = \prod_{i=1}^{n} H_i(f). \tag{2.72}$$

From Equation 2.44 we conclude further that the overall gain function is the product of the individual gains, that is,

$$G(f) = \prod_{i=1}^{n} G_i(f) \tag{2.73}$$

and that the overall phase shift is the sum of the individual phase shifts, or

$$\phi(f) = \sum_{i=1}^{n} \phi_i(f). \tag{2.74}$$

To find the output y(t) from the n systems in series it is necessary only to transform Equation 2.71 back to the time domain. Thus,

$$y(t) = \int_{-\infty}^{\infty} H_1(f) \cdot H_2(f) \ldots H_n(f) \cdot X(f) \, e^{i2\pi ft} \, df. \tag{2.75}$$

This approach to obtain y(t) is more straight forward than using the convolution integral approach wherein y(t) is found in terms of x(t) and the sequence of impulse response functions h_1, h_2, ..., h_n.

2.9 Ideal interpolation formula

The goal of the last section of this chapter is to show that there is a way to reconstruct an analog time series given only its digitally sampled version. As we might expect, there are important restrictions, but if they could be fully met the reconstruction would be exact. In practice, however, one of the two restrictions cannot be completely met, so only an approximate reconstruction is possible. The formula that performs the reconstruction is called the *ideal interpolation formula* and is in the form of a convolution integral (look ahead to Equation 2.89). Our objective is to develop this formula.

The first step is to derive the Fourier transform of an infinite train of unit impulse functions (refer to Section 2.5.1) given by

$$x^i(t) = \sum_{k=-\infty}^{\infty} \delta(t - k\Delta t), \quad -\infty \le t \le \infty \tag{2.76}$$

and shown in Figure 2.20. The sequence of vertical lines and arrows represents a train of delta functions (Section 2.4). The superscript *i* (unrelated to the imaginary unit i) is used throughout this section to indicate we are dealing with a generalized function, namely, the delta function. Because the impulses are periodic with period Δt, as delineated by the dashed line, we need only transform a finite portion of the signal, from $-\Delta t/2$ to $\Delta t/2$, rather than the infinite series. Following Equation 2.13, the Fourier transform of the delta function in the delineated section is

$$Q_m = \frac{1}{\Delta t} \int_{-\Delta t/2}^{\Delta t/2} \delta(t) \, e^{-i2\pi f_0 mt} \, dt, \quad -\infty \le m \le \infty \tag{2.77}$$

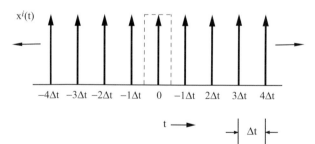

Figure 2.20 An infinite train of unit impulse functions. The area under each impulse has unit value.

where $Q_m = S'_m$ and $f_0 = 1/\Delta t = 1/T$. Using Equation 2.24, the Fourier amplitude at harmonic m reduces to

$$Q_m = \frac{1}{\Delta t}. \quad -\infty \leq m \leq \infty. \tag{2.78}$$

From Equation 2.14 and its periodic extension, the infinite train of delta functions can be alternatively written

$$x^i(t) = \sum_{m=-\infty}^{\infty} \frac{1}{\Delta t} e^{i2\pi f_0 mt}. \quad -\infty \leq t \leq \infty. \tag{2.79}$$

Using Equation 2.15 to obtain the Fourier transform of Equation 2.79 results in

$$X^i(f) = \sum_{m=-\infty}^{\infty} \int_{-\infty}^{\infty} \frac{1}{\Delta t} e^{-i2\pi(f-f_0 m)t} \, dt \tag{2.80}$$

which, from Equation 2.21, reduces to

$$X^i(f) = \frac{1}{\Delta t} \sum_{m=-\infty}^{\infty} \delta(f - f_0 m) = \frac{1}{\Delta t} \sum_{m=-\infty}^{\infty} \delta\left(f - \frac{m}{\Delta t}\right). \tag{2.81}$$

Equation 2.81 is shown in Figure 2.21. Function $X^i(f)$ is the amplitude density spectrum of an infinite train of unit impulses or delta functions, either of which is sometimes referred to as an "infinite Dirac comb" (after P.A.M. Dirac). The coefficient $1/\Delta t$ serves to scale the unit area that results from integration of each delta function.

In summary, we have

$$FT\left[x^i(t)\right] = FT\left[\sum_{k=-\infty}^{\infty} \delta(t - k\Delta t)\right]$$

$$= \frac{1}{\Delta t} \sum_{m=-\infty}^{\infty} \delta(f - mf_0) = \frac{1}{\Delta t} \sum_{m=-\infty}^{\infty} \delta\left(f - \frac{m}{\Delta t}\right) \tag{2.82}$$

$$= X^i(f).$$

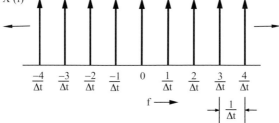

Figure 2.21 Amplitude density spectrum of an infinite train of unit impulse functions. The area under each delta function is $1/\Delta t$.

The Fourier transform of an infinite train of unit impulses in the time domain produces an infinite train of impulse functions in the frequency domain, each of which
has magnitude $1/\Delta t$ after integration. Using Equations 2.16 and 2.22, the inverse
Fourier transform of Equation 2.82 will lead directly to Equation 2.79, the original
train of unit impulse functions, $x^i(t)$.

The second step in obtaining the ideal interpolation formula is to apply Equation 2.82 to an infinite digitized time series. We begin by multiplying an infinite
analog time series, $s(t)$, by the infinite train of unit impulses, $x^i(t)$, then transforming
the product to obtain the amplitude density spectrum. Mathematically,

$$s^i(t) = s(t)\, x^i(t) \tag{2.83}$$

where $s^i(t)$ is the infinite train of digitally sampled values along the time axis. The
function $s^i(t)$ is itself an infinite train of impulse functions, each impulse function of
the form $s_k\delta(t - k\Delta t)$, where s_k is the magnitude of $s(t)$ at the sampled locations.
The magnitude of each s_k corresponds to the integral over $s_k\delta(t - k\Delta t)$ as given by
Equation 2.32. Figure 2.22 shows $s(t)$, $s^i(t)$, and s_k in which $s^i(t)$ is the same as $x^i(t)$
(not explicitly shown) except for the scale factor s_k. The Fourier transform of
Equation 2.83 can be written schematically as

$$FT\left[s^i(t)\right] = FT\left[s(t)\, x^i(t)\right]. \tag{2.84}$$

Recalling that the Fourier transform of a product is equivalent to the convolution
of the Fourier transform of each quantity in the product, we obtain the amplitude
density spectrum of the digitally sampled infinite time series, $s^i(t)$, that is,

$$S^i(f) = \sum_{m=-\infty}^{\infty} \int_{-\infty}^{\infty} S(f')\frac{1}{\Delta t}\,\delta\left(f - \frac{m}{\Delta t} - f'\right)df'$$

$$= \sum_{m=-\infty}^{\infty} \frac{1}{\Delta t}\, S\left(f - \frac{m}{\Delta t}\right) \tag{2.85}$$

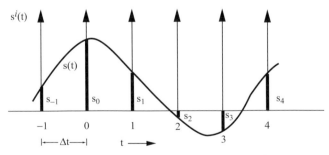

Figure 2.22 Continuous time series $s(t)$ sampled by $x^i(t)$ yielding $s^i(t)$.

where $S(f')$ is the infinite continuous transform of $s(t)$ and the Fourier transform of $x^i(t)$ is given by Equation 2.81. It is interesting to observe that while $s^i(t)$ is an infinite train of impulse functions, its Fourier transform is not. This is because the amplitudes of $s^i(t)$ are variable as determined by $s(t)$. If this were not the case and $s(t)$ were constant with time, then its transform would be a delta function (see Equation 2.21) and Equation 2.85 would be the convolution of an impulse function with a train of impulse functions. $S^i(f)$ is an aliased spectrum such that the amplitude density at any frequency is the sum of the amplitude densities at that frequency and all frequencies that are positive and negative integer multiples of $1/\Delta t$ away from that frequency.

In the third step, Equation 2.85 is transformed back to the time domain with the following restriction: $S(f)=0$ for $|f| >1/(2\Delta t)$; that is, the amplitude density spectrum is zero beyond the Nyquist frequency. With this restriction, $S^i(f)$ is simply the periodic version of $S(f)$. A hypothetical $S(f)$ and the resulting $S^i(f)$ are shown in Figure 2.23 in which the heavy line corresponds to the principal part of the complete aliased spectrum (as shown in Figure 1.28). Mathematically, we can multiply both sides of Equation 2.85 by the rectangular window, $W(f)$, where

$$W(f) = \begin{cases} \Delta t, & |f| \le \dfrac{1}{2\Delta t} \\ 0, & |f| > \dfrac{1}{2\Delta t} \end{cases} \qquad (2.86)$$

as seen by the dashed line in Figure 2.23.

Multiplying $S^i(f)$ by $W(f)$ results in a spectrum in which only the principal part is nonzero, yielding the amplitude density functions of the analog time series. Thus,

$$W(f)\, S^i(f) = S(f). \qquad (2.87)$$

Applying the inverse Fourier transform to the product, we have

$$FT^{-1}[W(f)\, S^i(f)] = FT^{-1}[S(f)]. \qquad (2.88)$$

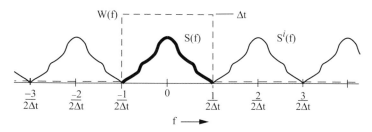

Figure 2.23 Amplitude density spectrum S(f), complete aliased spectrum $S^i(f)$, and rectangular window W(f).

The Fourier transform on the left side of Equation 2.88 becomes the convolution of the Fourier transforms of each term in the brackets, while the transform on the right side is simply s(t). Recalling that the Fourier transform of the rectangular function W(f) has the form of a diffraction function $(\sin z)/z$ (see Equations 2.19a and 2.19b), Equation 2.88 becomes

$$s(t) = \int_{-\infty}^{\infty} \frac{\sin(\pi u/\Delta t)}{\pi u/\Delta t} s^i(t-u)\, du = \int_{-\infty}^{\infty} s^i(u) \frac{\sin[\pi(t-u)/\Delta t]}{\pi(t-u)/\Delta t}\, du \quad (2.89)$$

where $s^i(u)$ is the infinite train of impulse functions with the amplitude of each impulse function after integration equal to the value of the sampled signal at intervals of Δt and zero elsewhere. The only contributions from the integration to s(t) result when the argument of $s^i(u)$ or $s^i(t-u)$ is nonzero – and nonzero values occur only at locations in time that are a multiple of Δt. The convolution process in the right-hand integral of Equation 2.89 is illustrated in Figure 2.24. For example, each nonzero value of $s^i(u)$ during integration (the magnitudes s_k in Figure 2.22) is multiplied by the appropriate value of the diffraction function centered at $t = t_1$; the completed integral for $t = t_1$ yields the value of $s(t = t_1)$. Figure 2.24 shows that $s(t_1)$ is slightly positive. Repeating this process for all t yields s(t). The same result follows from the left-hand convolution integral but it is more difficult to demonstrate because the diffraction function is symmetric whereas $s^i(t-u)$ is not. In application, the integrals in Equation 2.89 would be replaced by summations.

The same steps can be carried out in the ω-domain. Because $\omega_0 = 2\pi/\Delta t$, delta functions in the plot of $X^i(\omega)$ occur at integer multiples of $2\pi/\Delta t$. The relevant basic Fourier transform pair in the ω-domain is given by Equations 2.17 and 2.18 and other formulas in Section 2.4.

Assuming there is interest in reconstructing an original analog signal from its digital representation, the fidelity of the reconstruction will depend on the degree to which the two underlying restrictions in the above development are met. The restrictions were that the analog time series is band limited, that is, its amplitude

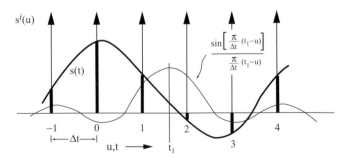

Figure 2.24 Application of the ideal interpolation formula. The diffraction function is convolved wth $s^i(u)$ to yield s(t).

density is zero beyond the Nyquist frequency, and that it has infinite record length. In practice, the former can be met exactly or quite closely if the amplitude density or variance beyond the Nyquist frequency is sufficiently small. For the latter restriction, it is, of course, not possible to have an infinite record. Nevertheless, even a moderately long time series can be quite satisfactory because the magnitude of the side lobes of the diffraction function decreases with distance from the main lobe. On the other hand, there will consequences near the ends of the time series. The situation is analogous to that discussed in Section 1.5.2 concerning spectrum windows. Near the middle of a periodogram the spectrum window could be accurately modeled as the square of the diffraction function (there in the frequency domain). This was not the case near the zero and Nyquist frequencies.

Finally, it should be noted that in addition to Equation 2.89 being referred to as the ideal interpolation formula it is known also as Whittaker's sampling formula and the cardinal interpolation formula (Jenkins and Watts, 1968). The diffraction function alone, $\sin(\pi u/\Delta t)/(\pi u/\Delta t)$, is sometimes referred to as the "ideal inter- polation function."

Problems

1 For a steady-state first-order linear system with input $x(t) = A \cos (2\pi ft)$, show that the output $y(t)$ intersects $x(t)$ at the extrema of $y(t)$.

2 Show, mathematically, that the general convolution integral

$$f_3(t) = \int_{-\infty}^{\infty} f_1(u) \, f_2(t-u) \, du$$

and the form

$$f_3(t) = \int_{-\infty}^{\infty} f_2(u) \, f_1(t-u) \, du$$

are equivalent.

3 Consider the following physical linear system:

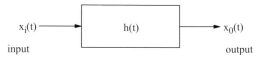

$x_i(t)$ ⟶ $h(t)$ ⟶ $x_0(t)$

input   output

(a) What are the two alternative equations for the convolution integral relating output x_0 to input x_i? Be sure to include the appropriate limits.

(b) Select one of the two equations and explain the connection between the limits on the integral and "physical realizability."

4 Write down the three principal properties of the delta function $\delta(t)$.

5 Recall that we can find the value of a function at a given point in time using the sifting property

$$\int_{-\infty}^{+\infty} \delta(t - t_0)\, s(t)\, dt = s(t_0).$$

Similarly, it is possible to define the m-th derivative of a delta function, namely, $\delta^{(m)}(t)$, that can be used to select the m-th derivative of a function at a given point. The formula is

$$\int_{-\infty}^{+\infty} \delta^{(m)}(t - t_0)\, s(t)\, dt = (-1)^m\, s^{(m)}(t_0), \quad m \geq 1.$$

Here, we will use this result to determine the frequency response function of a system that serves as a differentiator. In real life this could be a device that differentiates position with respect to time to obtain speed. The signal flow is:

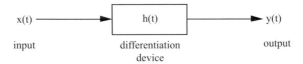

input   differentiation   output
                            device

The impulse response function is $h(u) = \delta^{(1)}(u)$. Derive the following:

(a) $H(f)$, the frequency response function.
(b) $G(f)$, the gain function.
(c) $\phi(f)$, the phase function.

6 A practical application of a first-order linear system is the resistance–capacitance circuit, or R–C filter, shown in the figure below. Here, $v_i(t)$ and $v_0(t)$ are the time varying input and output voltages, respectively.

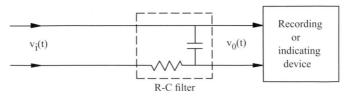

R-C filter

(a) Given that the product of the resistance, R (ohms), and capacitance, C (farads), is equal to the time constant (in seconds) of this one-parameter physical system, write down the differential equation that governs its performance.

(b) Determine the value of the time constant if $R = 1 \times 10^5$ ohms and $C = 1 \times 10^{-6}$ farads.

(c) Write down the impulse response function, h(t), for this system.

(d) Based on the illustration below, sketch h(t) indicating its value at time $t = 0$, where the heavy line with an arrow represents a unit impulse function (delta function).

(e) Write down the Fourier transform pair between H(ω) and h(t). Note that $\omega = 2\pi f$ and $d\omega = 2\pi df$.

(f) The expression for the voltage gain is

$$G(\omega) = \frac{V_0(\omega)}{V_i(\omega)} = \left[1 + (\omega RC)^2\right]^{-1/2}.$$

Compute the gain for values of $(\omega RC) = 0.01, 0.1, 1, 10, 100$, and 1000. Based on these values, sketch the gain function on a log-log graph with G(ω) on the y-axis and ωRC on the x-axis.

(g) Derive the mathematical expression for the slope of log $[G(\omega)]$ versus log (ωRC).

(h) What is the asymptotic value of this slope as ωRC becomes large? Make sure your plot of the gain function in (f) is in agreement with this value. (This kind of log-log plot is known as a Bode plot in electronic network analysis.)

7 Consider a propeller anemometer that is assumed to obey a first-order linear differential equation. Thus, we already know the system function (or impulse response function). An experiment is performed with the objective of determining the anemometer's time constant. The experiment involves placing the anemometer in a wind tunnel in which the speed is set at 2 m/s. The propeller is prevented from rotating until time $t = 0$, at which time it is released. The recorded data are given below, where $y(t)$ is the measured anemometer speed:

t (sec)	y(t) (m/s)
0	0.00
1	0.66
2	1.10
4	1.60
6	1.82
8	1.92
10	1.96

(a) Plot $y(t)$ on linear axes. Apply a smooth fit to the points (showing them) using, for example, a spline fit available in some graphics packages.

(b) Write down the convolution integral for the recorded anemometer speed that includes the formal expression for the step function, justifying your selection of limits on the integral. Then, carry out the integration to obtain an analytic expression for the output $y(t)$ for the given input.

(c) Obtain an analytic expression for the time constant. Using the above data and the analytic expression, obtain a value for the time constant.

(d) The time constant can be determined graphically given the expression for the ratio of the anemometer speed to the tunnel speed when t is equal to the time constant. Determine the numerical value of the ratio and then use this ratio and the graph in (a) to find the time constant.

(e) If the anemometer is placed in an atmospheric environment, what will happen to an input sinusoid at a frequency $f_1 = 0.2$ Hz? That is, determine the gain $G(f_1)$ and phase angle $\phi(f_1)$. Plot on a graph with linear axes an input sinusoid of amplitude 3 m/s (with, say, a mean wind speed of 10 m/s) and its associated output sinusoid. Show also the phase angle.

(f) At what angular frequency, ω', will the output variance of the sinusoid be one-half the input variance for a general first-order system? What is ω' for the conditions of this problem?

8 Given that $Y(\omega) = H(\omega) \, X(\omega)$ for a general linear system, take its inverse Fourier transform to show that $y(t) = h(t)^* x(t)$, where the asterisk notation has the usual meaning of convolution.

9 (a) Given a continuous stationary time series $x(t)$ that is digitally sampled with time interval Δt, thereby yielding x_t, under what two conditions can the original time series $x(t)$ be exactly reconstructed from x_t?

(b) Which of these conditions can be met in practice?

(c) Suppose you were given a realization of 100 values from a stationary process that met the condition in (b) and you reconstructed the time series using the ideal interpolation formula and the sample values. For which portion(s) of the realization would the reconstructed time series, $x_r(t)$, provide the best approximation to $x(t)$? For which portion(s) would it provide the poorest approximation to $x(t)$? Explain your reasoning.

10 Let us say that your trusted friend gives you a long digital time series that resulted from sampling (digitizing) an analog signal every 0.1 second. What is the highest frequency beyond which there can be no variance in the analog signal in order for you to use the ideal interpolation formula to reconstruct – reasonably well – the analog signal?

11 Given the following continuous signal flow where $h_1(t)$ and $h_2(t)$ are linear filters:

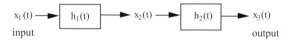

(a) Write down the convolution integral relating output $x_3(t)$ to input $x_1(t)$. Be careful with the dummy variables of integration.

(b) Let h_1 and h_2 be the impulse response functions for two first-order linear systems in which the time constant for h_1 is τ_1 and that for h_2 is τ_2. Write down the convolution integral for $x_3(t)$ that contains the specific expressions for these two impulse response functions.

(c) By finding $X_3(f)$ and then transforming it back to $x_3(t)$, derive the relationship

$$x_3(t) = \int_{-\infty}^{+\infty} \left\{ \left[1 + i(2\pi f \tau_1)\right] \left[1 + i(2\pi f \tau_2)\right] \right\}^{-1} X_1(f) \, e^{i2\pi ft} \, df$$

References

Brock, F.B. (1986) A nonlinear filter to remove impulse noise from meteorological data. *J. Atmos. Ocean. Tech.*, **3**, 51–58.

Cooper, G.R., and McGillem, C.D. (1967) *Methods of Signal and System Analysis*. Holt, Rinehart and Winston, New York.

Cooper, G.R., and McGillem, C.D. (1999) *Probabilistic Methods of Signal and System Analysis*, 3rd edn. Oxford University Press, New York.

Dirac, P.A.M. (1947) *Principles of Quantum Mechanics*, 3rd edn. Oxford Univ. Press, Oxford.

Jenkins, G.M., and Watts, D.G. (1968) *Spectral Analysis and its Applications*. Holden-Day, San Francisco, CA.

Koopmans, L.H. (1974) *The Spectral Analysis of Time Series*. Academic Press, New York.

Lumley, J.L. (1972) *Stochastic Tools in Turbulence*. Academic Press, New York.

3

Filtering data

In this chapter we investigate the properties of selected filters that can be applied to one-dimensional time series. To begin the investigation, we can think of the time series to be filtered as an input signal to which we apply a *weight function*, thereby yielding an output signal. The result of our action is embodied in the frequency response function that comprises a gain function and a phase function. The gain function tells us how the amplitude spectrum of the input will be modified by the weight function to yield the amplitude spectrum of the output. The phase function tells us how the phase spectrum of the input will be altered to yield the phase spectrum of the output; that is, it gives us the frequency-dependent phase angle changes, if any, that occur between the input and output signals. These functions and their analytical development were discussed in Chapter 2 (in particular, Equation 2.44) and provide the background needed in this chapter. Recall that the emphasis in Chapter 2 was on input–output relationships for a general linear physical system followed by applications to a system that obeyed a first-order linear differential equation and to an integrating device. What is different here is that we are dealing with data from a physical system that already have been collected or that are real-time data we wish to filter. The physical system *per se* is not of interest, only its output, which now becomes our input. In Chapter 2 we used the terms *system function* and *impulse response function* because we were dealing directly with physical systems. Here we use the term weight function in their place but carry over the term frequency response function.

An important aspect of this chapter is how to design a filter for application to digital data. By this is meant finding the number of weights and their values, that is, the weight function, needed to achieve a desired frequency response. We apply these design properties to a particular filter known as the Lanczos filter.

Time Series Analysis in Meteorology and Climatology: An Introduction, First Edition. Claude Duchon and Robert Hale.
© 2012 John Wiley & Sons, Ltd. Published 2012 by John Wiley & Sons, Ltd.

3.1 Recursive and nonrecursive filtering

There are two general types of data filters. A *recursive filter* is one in which the current output is related to the input and to the previous output; a *nonrecursive filter* is one in which the output is related only to the input. We begin our investigation by examining these two general types of filters for analog and digital data.

3.1.1 Analog data

When the input to the filtering process is an analog signal, we have

$$y(t) = \int_{-\infty}^{\infty} w(u)\, x(t-u)\, du + \int_{0}^{\infty} g(\tau) y\, (t-\tau)\, d\tau, \quad -\infty < t < \infty \quad (3.1)$$

in which $y(t) =$ output signal, $x(t) =$ input signal, $w(u) =$ weight function applied to the input signal, and $g(\tau) =$ weight function applied to the output signal.

The second convolution integral on the right-hand side accounts for Equation 3.1 being a recursive filter. As shown in Figure 3.1, a recursively filtered signal comprises the output from its weighted past (the recursive part) and the output from the weighted input (the nonrecursive part). If there is no recursion (no feedback), Equation 3.1 reduces to the familiar convolution integral

$$y(t) = \int_{-\infty}^{\infty} w(u)\, x(t-u)\, du. \quad (3.2)$$

Equation 3.2 is the general form of a nonrecursive filter.

3.1.2 Digital data

For the case when the input data to the system are digital

$$y_t = \sum_{k=-\infty}^{\infty} w_k\, x_{t-k} + \sum_{m=1}^{\infty} g_m\, y_{t-m}, \quad -\infty < t < \infty \quad (3.3)$$

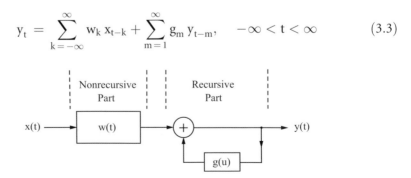

Figure 3.1   Schematic diagram of an analog recursive filter.

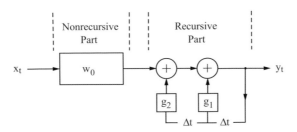

Figure 3.2 Schematic diagram of a digital recursive filter.

in which the notation parallels that for Equation 3.1. An example is

$$Y_t = w_0 x_t + g_1 y_{t-1} + g_2 y_{t-2}.$$ (3.4)

This example is shown schematically in Figure 3.2, where a single weight is used in the nonrecursive portion of the filter and the recursive portion of the output y_t is the sum of the output two time steps back multiplied by g_2 plus the output one time step back multiplied by g_1.

When $g_m = 0$, Equation 3.3 becomes the nonrecursive filter

$$Y_t = \sum_{k=-\infty}^{\infty} w_k x_{t-k}, \quad -\infty < t < \infty.$$ (3.5)

One application of recursive filters is in generating random processes, often for prediction purposes in science, engineering, and economics. We provide a simple illustration using Equation 3.4 and write it in the form of a prediction, one time step ahead. Thus,

$$Y_{t+1} = w_0 x_{t+1} + g_1 y_t + g_2 y_{t-1}.$$

If t is the current time, the next value of variable y (y_{t+1}) is predicted using the current value of y (y_t) and the previous value of y (y_{t-1}). The error of the prediction is the term $w_0 x_{t+1}$ because future values of the input are unknown. A similar calculation can be used to predict the value of y at time $t + 2$ given the predicted value y_{t+1} and current value y_t. You will have an opportunity to study the error of prediction for a simple statistical data model in problem 12 at the end of Chapter 4.

In the remainder of this chapter, our interest lies in nonrecursive filtering.

3.1.3 Low-pass, high-pass and band-pass filters

In nonrecursive data filtering it is common to refer to one of the three names given in the section title in order to identify which part of the frequency spectrum the Fourier amplitudes are to be retained or passed. Hence a *low-pass* filter is one in which the

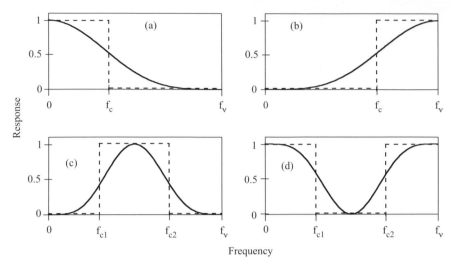

Figure 3.3   Ideal (dashed lines) and actual (solid lines) responses for (a) a low-pass filter, (b) a high-pass filter, (c) a band-pass filter, and (d) a band-stop filter. The quantity f_v is the Nyquist frequency and f_c is a 'cut-off' or 'cut-in' frequency.

amplitudes at the low frequency end of the Fourier spectrum, starting at the origin, are retained, while amplitudes at the high frequency end, toward the Nyquist frequency, are removed or suppressed. An example is shown in Figure 3.3a. A *high-pass* filter has the opposite result in the frequency domain, an example of which is shown in Figure 3.3b. A *band-pass* filter is one in which only amplitudes away from both the low and high frequency ends of the spectrum are retained as seen in Figure 3.3c. Its companion is the *band-stop* filter shown in Figure 3.3d, in which amplitudes in the interior of the spectrum are deleted. In each of the four examples in Figure 3.3 there is a dashed line and a solid line. The former represents what might be thought of as an *ideal* filter; that is, there is a 0-th order discontinuity or transition from unit response to zero response or vice versa. The frequency at which the discontinuity occurs is denoted by f_c for low- and high-pass filters and by f_{c1} and f_{c2} for band filters because two discontinuities are involved. Frequencies f_c for low-pass and f_{c2} and f_{c1} for band-pass and band-stop filters, respectively, are often referred to as the "cut-off" frequency (unit response to zero response), while f_c for high-pass and f_{c1} and f_{c2} for band-pass and band-stop filters, respectively, are appropriately referred to as the "cut-in" frequency (zero response to unit response). The solid lines are more appropriate to reality wherein a continuous transition from unit response to zero response, or the opposite, can be expected. In fact, none of the ideal response functions are achievable in practice.

Illustrations of the results of applying a high quality low-pass filter and a high quality high-pass filter to a time series are shown in Figure 3.4. Panel (a) shows a slowly varying noisy-looking data sequence y_t. Panel (b) is y_t after applying the

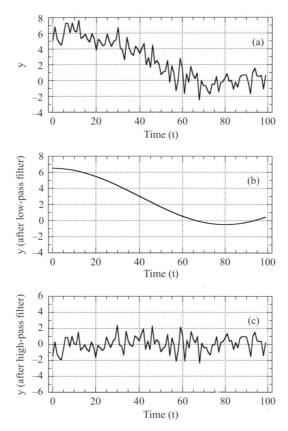

Figure 3.4 A time series of 100 values is shown in panel (a). Panel (b) is the same time series after applying a lowpass filter and panel (c) after applying a high-pass filter.

low-pass filter and panel (c) is y_t after applying the high-pass filter. The smoothly varying signal in Figure 3.4b reflects the trend visually evident in Figure 3.4a; with the trend removed, Figure 3.4c shows that time series y_t is distributed about zero. The high-frequency fluctuations in panels (a) and (c) can be easily matched.

Examples of specific simple low-pass and high-pass filters are described in Section 3.2. In the next section we discuss the connection between a low-pass filter and a high-pass filter and the mean value of the filtered data.

3.1.4 Preserving and removing the mean value of a time series

In applying a low-pass or band-stop filter to a data set, one wishes to *preserve*, as best one can, the mean of the time series. Similarly, in using a high-pass filter or band-pass filter, one desires to *remove* the mean of the time series. In this section we derive the criteria for meeting these objectives, first for an infinite analog time series, then for an infinite digital time series. Then we look at the application of these criteria to preserving or removing the mean for a time series of finite length, whether it is analog or digital.

3.1.4.1 Infinite analog time series

Consider the infinite input series $x(t)$ to which is applied the nonrecursive filter function $w(t)$. From Equation 3.2 the convolution integral has the form

$$y(t) = \int_{-L/2}^{L/2} w(u)\, x(t-u)\, du, \qquad -\infty \le t \le \infty \tag{3.6}$$

where L is the filter length and $y(t)$ the filtered output. The mean value of the output is obtained by taking the limit of the time-dependent average of $y(t)$ according to

$$\bar{y} = \lim_{T \to \infty} \frac{1}{T} \int_{-T/2}^{T/2} y(t)\, dt$$

$$= \lim_{T \to \infty} \frac{1}{T} \int_{-T/2}^{T/2} \int_{-L/2}^{L/2} w(u)\, x(t-u)\, du\, dt$$

$$\bar{y} = \int_{-L/2}^{L/2} w(u) \left[\lim_{T \to \infty} \frac{1}{T} \int_{-T/2}^{T/2} x(t-u)\, dt \right] du \tag{3.7}$$

where T is the length of the averaging interval. In the limit as the averaging interval tends to infinity, the term in the brackets becomes $\bar{x}$, so that

$$\bar{y} = \bar{x} \int_{-L/2}^{L/2} w(u)\, du. \tag{3.8}$$

This result implies that

$$\int_{-L/2}^{L/2} w(u)\, du = \begin{cases} 1, & \text{mean preserved} \\ 0, & \text{mean removed} \end{cases}. \tag{3.9}$$

Thus, when the integral of the weight function is one, the mean after filtering is the same as the mean before filtering. When the integral is zero, the mean after filtering is zero regardless of the mean before filtering.

3.1.4.2 Infinite digital time series

For digital data we apply the nonrecursive filter weight function w_k to the time series x_t. From Equation 3.5 the expression for the convolution summation is

$$y_t = \sum_{k=-n}^{n} w_k \, x_{t-k}, \quad t = 0, \pm 1, \pm 2, \cdots \quad (3.10)$$

where $(2n + 1)$ is the length of filter w_k and y_t is the filtered output. By analogy with the analog case above, the mean value of the output is given by

$$\bar{y} = \lim_{N \to \infty} \frac{1}{N} \sum_{t=-(N-1)/2}^{(N-1)/2} y_t$$

$$= \lim_{N \to \infty} \frac{1}{N} \sum_{t=-(N-1)/2}^{(N-1)/2} \sum_{k=-n}^{n} w_k \, x_{t-k}$$

$$\bar{y} = \sum_{k=-n}^{n} w_k \left[\lim_{N \to \infty} \frac{1}{N} \sum_{t=-(N-1)/2}^{(N-1)/2} x_{t-k} \right] \quad (3.11)$$

where N is the number of data in the averaging interval, and N is odd. In the limit as the averaging length approaches infinity, the term in brackets becomes $\bar{x}$. Thus,

$$\bar{y} = \bar{x} \sum_{k=-n}^{n} w_k \quad (3.12)$$

implying that

$$\sum_{k=-n}^{n} w_k = \begin{cases} 1, & \text{mean preserved} \\ 0, & \text{mean removed} \end{cases}. \quad (3.13)$$

Therefore, identical to the infinite analog case, when the sum of the weights is one, the mean after filtering is the same as the mean before filtering. When the sum of the weights is zero, the mean after filtering is zero. Note that Equations 3.8 and 3.12 show no restriction on the value of individual weights, only their sum.

3.1.4.3 Finite time series

In practice, the length of the time series is always finite, regardless of whether the data are analog or digital. In Figure 3.5 the heavy line is a time series of length

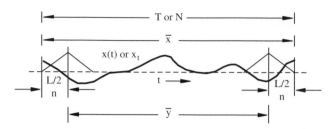

Figure 3.5 The length of a filtered time series relative to the length of the original finite time series is reduced by the length of the filter.

T (analog) or N (digital) and a triangular filter (discussed in Section 3.2.2) of corresponding length L or (2n+1) is shown at both ends of the time series.

Filtered values y(t) or y_t cannot be calculated any closer than one-half the filter length, L/2 or n, to either end of the time series of x(t) or x_t. Any attempt to compute a filtered time series beyond L/2 or n at either end of the time series in the illustration will yield an incorrect filtered time series. Thus, when the objective is to have $\bar{y}$ equal to either $\bar{x}$ or 0, this will occur, in general, only in the limit as T or N tend to infinity. Good approximations to $\bar{y}=\bar{x}$ or $\bar{y}=0$, however, can be realized when the length of the filter is small relative to the length of the data record, that is, when the "end effect" is small.

3.2 Commonly used digital nonrecursive filters

In this section we examine the weight function and resulting frequency response function of several low-pass filters and one high-pass filter that are often used because their weights are easy to calculate. These filters belong to the category of one-parameter filters; that is, the only quantity that can be varied is the number of weights.

In Chapter 2 we established the relationship between the frequency response function H(f) and the system or unit impulse response h(t). This was done in Section 2.6 for analog data and resulted in the Fourier transform

$$H(f) = \int_{-\infty}^{\infty} h(t)\,\exp(-i2\pi ft)\,dt. \qquad (2.43)$$

Before we investigate the properties of easy-to-apply filters, we need to use Equation 2.43 to determine the formula for the transform where the system function is a digital weight function rather than a general analog function. As opposed to integration, we will need a summation over the weight function with the exponential term occurring only at the instances in time of the digital weights. Making use of the δ-function

(Section 2.4) to isolate these instances, we have

$$H(f) = \sum_{k=-\infty}^{\infty} w_k \int_{-\infty}^{\infty} \delta(t - k\Delta t) \exp(-i2\pi ft)\, dt, \quad k = 0, \pm1 \pm 2, \ldots$$

which reduces to

$$H(f) = \sum_{k=-\infty}^{\infty} w_k \exp(-i2\pi fk\Delta t). \tag{3.14}$$

H(f) contains all the information about amplitude and phase angle changes that result from filtering a time series with weight function w_k.

3.2.1 Running mean filter

The weight function for the running mean or rectangular filter is

$$w_k = \begin{cases} 1/(2n+1), & 0 \le |k| \le n \\ 0, & |k| > n \end{cases} \tag{3.15}$$

and is shown in Figure 3.6 for the total number of weights $(2n+1)=11$.

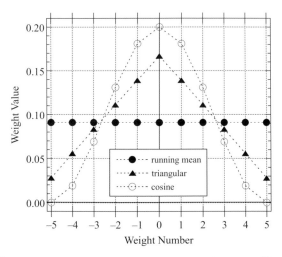

Figure 3.6 Weight value versus weight number for the running mean filter in Equation 3.15, the triangular filter in Equation 3.19, and the cosine filter in Equation 3.23. The number of weights for each filter is $(2n+1)=11$.

From Equation 3.14 we have

$$H(f) = \frac{1}{2n+1} \sum_{k=-n}^{n} \exp(-i2\pi f k \Delta t).$$

Using Equation 1.B.4 this becomes

$$H(f) = \frac{\sin[\pi f(2n+1)\Delta t]}{(2n+1)\sin(\pi f \Delta t)}. \tag{3.16}$$

According to Equation 3.16, the first zero crossing (i.e., the argument in the numerator is equal to π) occurs at the frequency that is inversely proportional to the length of the filter; subsequent crossings occur at multiples of this frequency. As an example, let's use the weight function in Figure 3.6 and let the sampling or data interval Δt equal one week. Figure 3.7 shows that the first zero crossing is at a frequency of 0.091 cycles/week. The second zero crossing is at a frequency of 0.182 cycles/week or a period of 5.5 weeks, and so on. Although the running mean is easy to apply, it has large negative and positive side lobes. Waveforms at frequencies with negative responses are inverted in the filtered data relative to the original data. Figure 3.7 also shows the frequency response functions for two other simple filters that will be discussed shortly.

For sufficiently large n, Equation 3.16 can be approximated by

$$H(f) = \frac{\sin(2\pi f n \Delta t)}{2\pi f n \Delta t}. \tag{3.17}$$

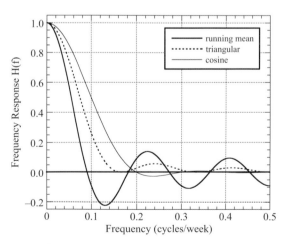

Figure 3.7 The response functions associated with the weight functions in Figure 3.6. The respective equations for the response functions are Equation 3.16, Equation 3.20, and Equation 3.26. The sampling interval Δt is one week.

Equation 3.17 has an absolute error of less than 0.05 for $n > 10$. For $n < 10$, Equation 3.16 should be used (as in Figure 3.7). We notice that Equation 3.17 is a diffraction function. This is not surprising because we saw in section 2.4 that the Fourier transform of an analog rectangular function (Equation 2.19) is a diffraction function.

If we replace the approximate filter length $2n\Delta t$ by L and frequency f by $1/T$, where T is the period, Equation 3.17 becomes

$$H(f) = \frac{\sin(\pi L/T)}{\pi L/T}. \tag{3.18}$$

Equation 3.18 is plotted in Figure 3.8. The zero crossings are located at $L/T = k = 1$, 2, ..., when the numerator in Equation 3.18 is zero. Each integer k is the number of complete cycles in the length of the running mean filter. Averaging an integer number of sinusoids always yields zero. Figure 3.8 provides a general way to assess the frequency response of simple filters as a result of altering the number of weights, data interval, and frequency.

Each of the frequency response functions Equations 3.16–3.18 could be placed in an alternative form involving gain and phase functions following Equations 2.43 and 2.44. Thus the absolute values of Equations 3.16–3.18 become their respective gain functions G(f). The phase function of each is the arctangent of the ratio of the imaginary part of H(f) to its real part. The imaginary part is zero because the running mean as defined above is an even function; the real part is nonzero and varies in algebraic sign with frequency. Furthermore, the angle defined by the arctangent of

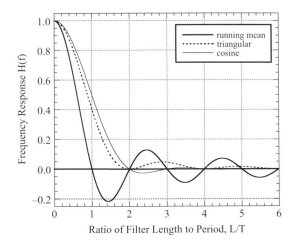

Figure 3.8   Approximate frequency response functions for a running mean or rectangular filter (Equation 3.18) for $n > 10$, a triangular filter (Equation 3.22) for $n > 15$, and a cosine filter (Equation 3.27) for $n > 1$. The number of weights is $2n+1$.

the ratio is either 0 or 180 degrees, as shown in Figures 3.7 and 3.8. Positive response corresponds to 0 degrees and negative response to 180 degrees.

3.2.2 Triangular filter

The weight function for the triangular filter can be derived from convolving two identical running mean filters. The derivation is carried out in Appendix 3.A with the result that

$$w_k = \begin{cases} \dfrac{1}{n+1}\left(1 - \dfrac{|k|}{n+1}\right), & 0 \le |k| \le n \\ 0, & |k| > n \end{cases}. \tag{3.19}$$

The weight sequence is plotted in Figure 3.6 for $2n+1 = 11$ weights. The response is derived in Appendix 3.B where it is shown that

$$H(f) = \frac{\sin^2[\pi f(n+1)\Delta t]}{(n+1)^2\sin^2(\pi f\Delta t)}. \tag{3.20}$$

The response function Equation 3.20 for 11 weights is shown in Figure 3.7. With $\Delta t = 1$ week, the first zero crossing is at 0.167 cycles/week, nearly double that for the running mean. At the same time, the side lobes are reduced relative to the running mean and always positive. Thus there is a trade-off between the rate of descent of the frequency response from the frequency origin and the absolute amplitude of the side lobes when both filters have the same number of weights.

Similar to Equation 3.16, an approximate form of Equation 3.20 is

$$H(f) = \frac{\sin^2(\pi fn\Delta t)}{(\pi fn\Delta t)^2} \tag{3.21}$$

for which the absolute error is less than 0.05 for $n > 15$. If we again let $L = 2n\Delta t$ and $f = 1/T$, Equation 3.21 becomes

$$H(f) = \frac{\sin^2[\pi L/(2T)]}{[\pi L/(2T)]^2}. \tag{3.22}$$

Equation 3.22 is plotted in Figure 3.8, in which the zero crossings are at $L/T = 2k$, $k = 1, 2, \ldots$. Because of the triangular shape of the weight function, two complete cycles of a sinusoid, or multiples thereof, over the filter length are required to yield zero response after convolution. A useful exercise is to show this graphically.

3.2.3 Cosine filter

Another easy-to-use filter is the cosine or raised-cosine or von Hann filter (Hamming, 1977, pp. 88–90). As with the running mean and triangular filters, it has one parameter, the number of weights. It is defined by

$$
w'_k = \begin{cases} \dfrac{1 + \cos(\pi k/n)}{2}, & |k| \le n \\ 0, & |k| > n \end{cases}. \tag{3.23}
$$

Because the sum of the $2n + 1$ weights must equal one, using Appendix 1.B we find that the expression for the standardized weight function is

$$
w_k = \begin{cases} \dfrac{1 + \cos(\pi k/n)}{2n}, & |k| \le n \\ 0, & |k| > n \end{cases} \tag{3.24}
$$

which is shown in Figure 3.6 for $2n + 1 = 11$ weights. We notice that the two end-weights are zero so that, effectively, they contribute nothing to a filtered value. That is, there are $2n - 1$ "working weights."

The derivation of the response function is initiated by

$$
\begin{aligned}
H(f) &= \sum_{k=-n}^{n} w_k \exp(-i2\pi f k \Delta t) \\
&= \frac{1}{(4n)} \sum_{k=-n}^{n} [\exp(i\pi k/n) + 2 + \exp(-i\pi k/n)] \exp(-i2\pi\, fk\,\Delta t).
\end{aligned}
$$

The completion of the derivation is lengthy and is given in the previous reference. The result is

$$
H(f) = \frac{\sin(2\pi n f \Delta t) \cos(\pi f \Delta t)}{2n \sin(\pi f \Delta t)} \left[\frac{1}{1 - \left[\frac{\sin(\pi f \Delta t)}{\sin(\pi/(2n))}\right]^2} \right] \tag{3.25}
$$

which can be approximated by

$$
H(f) = \frac{\sin(2\pi n f \Delta t)}{2\pi n f \Delta t} \left[\frac{1}{1 - (2n f \Delta t)^2} \right]. \tag{3.26}
$$

The absolute error of Equation 3.2.6 is less than 0.03 for $n > 1$. Equation 3.26 for $2n + 1 = 11$ weights is plotted in Figure 3.7, which shows that the main lobe of the cosine filter is somewhat broader than that of the triangular filter while the side lobes are reduced in amplitude and are alternately positive and negative.

If we let $L = 2n\Delta t$ and $f = 1/T$ as in the previous filters, Equation 3.26 becomes

$$H(f) = \frac{\sin(\pi L/T)}{\pi L/T}\left[\frac{1}{1 - (L/T)^2}\right]. \tag{3.27}$$

Equation 3.27 is plotted in Figure 3.8. The zero crossings coincide with the second and subsequent zero crossings of the running mean filter. Based on Figures 3.7 and 3.8 for these three simple filters, we conclude that, for the same number of weights, the smoother the weight function, the lower the side lobes in the frequency response function and the less rapid the descent in the main lobe.

3.2.4 Difference filter

The fourth and last one-parameter filter we examine is a high-pass difference filter. Among the four filters, it is the only one that has a complex response function $H(f)$. It is complex because its weight function is not symmetric about the time origin. The previous filters we studied each had a symmetric weight function, resulting in real response functions.

The high-pass difference filter we investigate has weight function

$$w_0 = 1/2, \quad w_1 = -1/2; \quad w_k = 0 \quad \text{elsewhere.} \tag{3.28}$$

The sum of the weights is zero so that the mean of a filtered time series will be removed. The formula for convolution is, from Equation 3.5,

$$y_t = (x_t - x_{t-1})/2. \tag{3.29}$$

Since the filter is asymmetric, there will be nonzero phase angle differences between the input and the output. The response function is

$$H(f) = \sum_{k=0}^{1} w_k \exp(-i2\pi f k \Delta t)$$

which reduces to

$$H(f) = \frac{1}{2}[1 - \exp(-i2\pi f \Delta t)]. \tag{3.30}$$

After extracting $\exp(-i\pi f\Delta t)$ from the right-hand side, we obtain

$$H(f) = i\sin(\pi f\Delta t)\exp(-i\pi f\Delta t)$$

or

$$H(f) = \sin(\pi f\Delta t)\exp\left[-i\pi\left(f - \frac{1}{2\Delta t}\right)\Delta t\right]. \qquad (3.31)$$

From Equation 2.44

$$H(f) = G(f)\exp[i\phi(f)]$$

so that

$$G(f) = |\sin(\pi f\Delta t)| \qquad (3.32)$$

and

$$\phi(f) = \begin{cases} -\pi\left(f - \dfrac{1}{2\Delta t}\right)\Delta t, & 0 \le f \le 1/(2\Delta t) \\[2mm] -\pi\left(f + \dfrac{1}{2\Delta t}\right)\Delta t, & -1/(2\Delta t) \le f < 0 \end{cases}. \qquad (3.33)$$

Because $G(f)$ is always positive, Equation 3.33 validates Equation 3.31 by taking into account that $\sin(\pi f\Delta t)$ is negative for $-1/(2\Delta t) \le f < 0$. The gain and phase functions for $w_0 = {}^1/_2$ and $w_1 = -{}^1/_2$ are shown in Figures 3.9 and 3.10 for positive frequency. As can be seen in Figure 3.9, the difference filter is a high-pass filter,

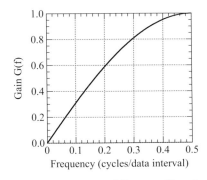

Figure 3.9   Gain function for the difference filter in Equation 3.29.

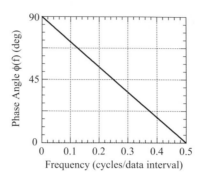

Figure 3.10   Phase function for the difference filter in Equation 3.29.

eliminating the mean and passing unchanged the amplitude at the Nyquist frequency. Figure 3.10 shows that as the frequency of a sinusoid approaches zero (wavelength tends to infinity), digital differencing is similar to taking the derivative of a sine function, resulting in a cosine function and a phase difference of 90°. At the Nyquist frequency, the filtering amounts to differencing successive maximum and minimum values which leaves the sinusoid unchanged. Equation 3.33 shows that there is a discontinuity in ϕ(f) at the frequency origin where its value is +90° as the origin is approached from the positive side and −90° as the origin is approached from the negative side. Phase function ϕ(f) increases linearly from −90 to 0° as the frequency decreases from −0 cycles per data interval to the negative Nyquist frequency.

3.2.5 Relationship between high-pass and low-pass filters

As noted in Section 3.1.4, when the sum of the filter weights is one, the mean of the time series is preserved, and when the sum of the weights is zero, the mean is removed. The respective frequency response functions are low-pass, with $H(f=0) = 1$, and high-pass, with $H(f=0) = 0$. Intuitively, we should expect that a high-pass filter $H'(f)$ could be derived from a low-pass filter $H(f)$ by subtraction according to

$$H'(f) = 1 - H(f). \tag{3.34}$$

That is, where the response was high in a low-pass filter it will be low in the high-pass filter, and vice versa. Using Equation 3.14 and assuming a finite number of weights, the frequency response function for the high-pass filter (Equation 3.34) becomes

$$H'(f) = 1 - w_0 - \sum_{k=-m}^{-1} w_k\, e^{-i2\pi fk\Delta t} - \sum_{k=1}^{n} w_k\, e^{-i2\pi fk\Delta t} \tag{3.35}$$

in which allowance has been made for an asymmetric filter. The weights for the high-pass filter are then

$$w_0' = 1 - w_0 \quad \text{and} \quad w_k' = -w_k, \quad k \neq 0. \tag{3.36}$$

When the filter is symmetric, $m = n$, and the high-pass frequency response function (Equation 3.34) reduces to

$$H'(f) = w_0' + 2 \sum_{k=1}^{n} w_k' \cos(2\pi f k \Delta t) \tag{3.37}$$

where we see that $H'(f)$ has no imaginary component since the sine term sums to zero. This form will be used again in Section 3.4.3 in discussing filter design.

In this section we examined the conversion of a low-pass filter to a high-pass filter, the result of which was the set of Equations 3.34–3.37. We could just as well have applied this equation set to the conversion of a high-pass filter to a low-pass filter. Moreover, these same equations apply to conversion of a band-pass filter to a stop-band filter and vice versa.

3.3 Filter design

In the previous section we examined a selection of commonly used filters and their response functions. A specific set of weights resulted in a specific response function. In this section we investigate the reverse problem. Given a desired response function, what is the required set of weights? Thus, the first step in designing a digital filter is to decide the shape of the frequency response function. The second step is to Fourier transform this function to the time domain to obtain the weight function or digital filter weights. Typically, to achieve the desired response function requires a greater number of weights than are practical, so the weight function must be truncated. The third step is to Fourier transform the truncated weight function back to the frequency domain to get the *actual response function*.

It is useful to think of the frequency response function as an analog periodic data series with frequency instead of time as the independent variable and whose period is $1/\Delta t$, the frequency interval between the Nyquist frequencies $-f_v$ and f_v. That we can view the frequency response function as a periodic function is because it comprises the principal part of the complete aliased spectrum. In this context, we represent the response function by a Fourier synthesis whose Fourier coefficients become the filter weights in the time domain.

Thus, by steps 1 and 2 we have, following Equations 2.13 and 2.14,

$$w_k = \left(\frac{1}{2f_v}\right) \int_{-f_v}^{f_v} H(f) \exp(-i2\pi f k \Delta t) \, df, \quad -\infty \leq k \leq \infty \tag{3.38}$$

and

$$H(f) = \sum_{k=-\infty}^{\infty} w_k \exp(i2\pi fk\Delta t)$$

$$= \sum_{k=-\infty}^{\infty} w_k \cos(2\pi fk\Delta t) + i \sum_{k=-\infty}^{\infty} w_k \sin(2\pi\, fk\Delta t). \tag{3.39}$$

In the framework of filter design, $H(f)$ can be appropriately called the *design response function*; as in previous sections, w_k is the weight function. Comparing Equations 3.38 and 3.39 with Equations 2.13 and 2.14, respectively, we see that w_k corresponds to S'_m, $2f_v$ to T, $H(f)$ to $x(t)$, df to dt, and $f\Delta t$ to t/T in the exponent.

After parsimonious truncation of the weight function, the filtered series takes the usual form of digital convolution (see Equation 3.5):

$$y_t = \sum_{k=-m}^{n} w_k x_{t-k}, \quad -\infty \le t \le \infty. \tag{3.40}$$

What is meant by "parsimonious" here is that the number of weights should be small compared to the length of the time series but large enough to provide good fidelity of the response function. The actual values of m and n are decided by the investigator.

Next, we anticipate the results of applying a Fourier transform to a convolution integral (or summation here) that we found in Section 2.7. Accordingly, we Fourier transform Equation 3.40 to the frequency domain to obtain the amplitude spectra of the filtered series y_t, the original time series x_t, and the weight function w_k. Thus,

$$Y(f) = \sum_{t=-\infty}^{\infty} \left[\sum_{k=-m}^{n} w_k x_{t-k} \right] \exp(i2\pi ft\Delta t)$$

$$= \sum_{k=-m}^{n} w_k \left[\sum_{t=-\infty}^{\infty} x_{t-k} \exp(i2\pi ft\Delta t) \right] \tag{3.41}$$

$$= \sum_{k=-m}^{n} w_k \exp(i2\pi fk\Delta t) \left[\sum_{t=-\infty}^{\infty} x_{t-k} \exp[i2\pi f(t-k)\Delta t] \right]$$

which reduces to

$$Y(f) = H_{m,n}(f)\, X(f). \tag{3.42}$$

Except for the interchange of variables, Equations 3.41 and 3.42 are similar to Equations 2.58 and 2.59 for analog data.

The actual response function

$$H_{m,n}(f) = \sum_{k=-m}^{n} w_k \exp(i2\pi fk\Delta t) \tag{3.43}$$

can be expanded into its real and imaginary parts as

$$H_{m,n}(f) = \sum_{k=-m}^{n} w_k \cos(2\pi fk\Delta t) + i \sum_{k=-m}^{n} w_k \sin(2\pi fk\Delta t). \tag{3.44}$$

Equation 3.44 can be written also

$$H_{m,n}(f) = \frac{Y(f)}{X(f)} \tag{3.45}$$

in which form it is the same as Equation 2.60 for analog time series and expresses the ratio of the output to the input complex amplitudes as a function of frequency f. Similar to Equation 2.44, the response function $H_{m,n}(f)$ can be written as the product of a gain function and an exponential term involving the phase function.

 If the design response function $H(f)$ in Equation 3.39 is real and symmetric about $f = 0$, then the weight function w_k is real and symmetric about $k = 0$. Only cosines are needed in the synthesis of $H(f)$. The same can be said of $H_{m,n}(f)$ if $m = n$. In fact, we will limit our investigation to symmetric filters with the result that their response functions are always real and $H_{m,n}(f) = H_n(f)$. Because the phase angle will be either $0°$ or $180°$ (as in the running mean), we will usually not use the terms gain function (or gain) or phase function (or phase), but instead, frequency response function or response function or, simply, response.

3.4 Lanczos filtering

In this section, we describe a Fourier filtering method called Lanczos filtering. Its principal feature is the use of "sigma factors," which significantly reduce the amplitude of the Gibbs phenomenon, an oscillation that occurs when a function is approximated by a partial sum of a Fourier synthesis. The Gibbs phenomenon or oscillation can result in considerable error in the vicinity of the discontinuity in an ideal response function, for example, the cut-off f_c in Figure 3.3a. The filter is a two-parameter symmetric filter, one parameter being the number of weights, the other the cut-off frequency. In the one-parameter filters in Section 3.2 the number of weights completely determined the frequency response function. Here we have an additional degree of flexibility. Using these parameters as entries to a pair of graphs, the frequency response function can be estimated. The simplicity of calculating the weights and the adequate response for many needs make Lanczos filtering an attractive filtering method.

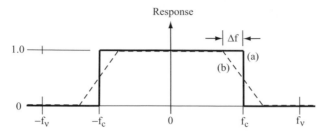

Figure 3.11   (a) An ideal low-pass response function with cut-off frequency f_c. (b) The smoothed ideal response function given by Equation 3.49. The transition band is $2\Delta f = 2f_v/n$ (Duchon, 1979).

3.4.1 Mathematical development

The material provided in this and the following two sections is based on a paper by Duchon (1979). Application of Lanczos filtering to data sets in two dimensions is also discussed in this paper. We begin by considering an ideal low-pass response function shown by the solid line in Figure 3.11 (or dashed line in Figure 3.3a) where f_c and f_v are the cut-off and Nyquist frequencies, respectively. Mathematically, we have

$$H(f) = \begin{cases} 1, & |f| \le f_c \\ 0, & |f| > f_c \end{cases}. \tag{3.46}$$

Using Equation 3.38, we Fourier transform H(f) to get its weight function

$$w_k = \left(\frac{1}{2f_v}\right) \int_{-f_c}^{f_c} \exp(-i2\pi fk\Delta t)\, df = \frac{\sin(2\pi f_c k\Delta t)}{2\pi f_v k\Delta t}, \quad -\infty \le k \le \infty. \tag{3.47}$$

In application, we have to limit the number of weights. Lanczos (1956, p. 219) showed that when the finite set of weights is transformed back to the frequency domain by Equation 3.43 where $m = n$, the departure from the ideal response function has the form of a "modulated carrier wave." The carrier frequency is equal to the frequency of the first term neglected in the Fourier synthesis and its amplitude contributes significantly to the Gibbs oscillation. Thus, as proposed by Lanczos, the carrier frequency should be filtered. This can be done using a "sigma factor." An example of Gibbs oscillation and its suppression is shown in Figure 3.12 and described in the next section.

   In the mathematical development in this section, a nonstandard formula for the total number of weights in a filter is employed. The standard formula is $2n + 1$, where n is the number of weights to the right and left of the central weight. The formula for the total number of weights used in Lanczos filtering is $2n-1$. For example, if there are a total of seven weights, the central weight (w_0) and three weights to the right (w_1, w_2, w_3) and three to the left (w_{-1}, w_{-2}, w_{-3}), $n = 4$.

The reason for choosing the nonstandard formula is because n is equal to the first term truncated in synthesizing H(f) and corresponds to n cycles over $2f_v$. In this example, the carrier frequency is $n = 4$ cycles over $2f_v$.

Therefore, to suppress the Gibbs oscillation for the general case of using $2n-1$ weights, we convolve the ideal response function Equation 3.46 with the rectangular function

$$h(f) = \begin{cases} \dfrac{n}{2f_v}, & |f| \leq f_v/n \\ 0, & |f| > f_v/n \end{cases}. \tag{3.48}$$

The width of rectangular filter $h(f)$ corresponds to the width of one cycle of a Gibbs oscillation and would result in its complete suppression were there no modulation. Due to the modulation (we no longer have a pure sinusoid) we can anticipate some residual oscillation after filtering. The area under $h(f)$ has unit value so that the mean of the function being filtered is unchanged.

The smoothed version of $H(f)$ is produced by the convolution integral

$$\overline{H}(f) = \int_{-f_v/n}^{f_v/n} h(g)\, H(f - g)\, dg$$

$$= \left(\frac{n}{2f_v}\right) \int_{-f_v/n}^{f_v/n} H(f - g)\, dg \tag{3.49}$$

where $h(g) = h(f)$. $\overline{H}(f)$ is shown by dashed curve (b) in Figure 3.11. We note that, as expected, the width of the frequency band from unit response to zero response is the same as the width of the nonzero part of $h(f)$, that is, $2\Delta f = 2f_v/n$. The narrow band from unit to zero response or vice versa is commonly referred to as the *transition band*.

Using Equation 3.38 again, the Fourier transform of Equation 3.49 yields the smoothed weight function

$$\overline{w}_k = \left(\frac{1}{2f_v}\right) \int_{-f_v}^{f_v} \overline{H}(f)\, \exp(-i2\pi f k \Delta t)\, df, \qquad -\infty \leq k \leq \infty. \tag{3.50}$$

For the case of a partial sum of Fourier synthesis we have, following Equation 3.43,

$$\overline{H}_n(f) = \sum_{k=-(n-1)}^{n-1} \overline{w}_k\, \exp(i2\pi f k \Delta t)$$

$$= \overline{w}_0 + 2 \sum_{k=1}^{n-1} \overline{w}_k\, \cos(2\pi f k \Delta t). \tag{3.51}$$

Equation 3.51 shows the actual response function as a consequence of smoothing (convolving) the ideal response function with a rectangular function that is tuned to the number of weights used, $(2n-1)$, to minimize the Gibbs oscillation. Now we need to find the values of the weights.

The result of substituting Equation 3.49 into Equation 3.50 and interchanging the order of integration is

$$\overline{w}_k = \left(\frac{n}{2f_v}\right) \int_{-f_v/n}^{f_v/n} \exp(-i2\pi gk\Delta t)$$

$$\times \left[\left(\frac{1}{2f_v}\right) \int_{-f_v}^{f_v} H(f-g) \exp[-i2\pi(f-g)k\Delta t] \, d(f-g)\right] dg \quad (3.52)$$

where the term in brackets is weight function w_k in Equation 3.47. Because $H(f)$ is periodic (Section 3.3) with period $2f_v$, Equation 3.52 reduces to

$$
\begin{aligned}
\overline{w}_k &= w_k \frac{\sin(2\pi k f_v \Delta t/n)}{2\pi k f_v \Delta t/n} \\
&= \frac{\sin(2\pi f_c k \Delta t)}{2\pi f_v k \Delta t} \frac{\sin(2\pi k f_v \Delta t/n)}{2\pi k f_v \Delta t/n}.
\end{aligned}
\quad (3.53)
$$

If we let $\Delta t = 1$ and $f_v = 0.5$ so that f has units of cycles/data interval, then

$$\overline{w}_k = \frac{\sin(2\pi f_c k)}{\pi k} \frac{\sin(\pi k/n)}{\pi k/n}, \quad k = -(n-1), \ldots, 0, \ldots, (n-1). \quad (3.54)$$

We see that the weight function for the smoothed response function is the product of that for the ideal response function and the term

$$\sigma = \frac{\sin(\pi k/n)}{\pi k/n} \quad (3.55)$$

called the "sigma factor" by Lanczos. A second smoothing of the ideal response function (Equation 3.46) with Equation 3.47 yields σ^2 in Equation 3.54 and subsequent smoothing by a corresponding increase in power of the sigma factor. Typically, only one smoothing is performed.

There is one caveat in the above derivation. In carrying out the integration of the interior integral on the right-hand side of Equation 3.52 it was assumed that $\overline{H}(f)$ in Equation 3.49 or the dashed line in Figure 3.11 became zero before $\pm f_v$. Mathematically, this can be expressed as $f_c + f_v/n < f_v$. If this criterion is not met, then Equation 3.53 is incorrect. The practical result is that the response function for a low-pass filter will never pass through zero.

3.4.2 Results

Curve (a) in Figure 3.12 is an ideal response function in which the cut-off frequency f_c is 0.2 cycles/data interval; curve (b) is the response function computed from Equation 3.45 in which there are $2n-1 = 19$ weights computed from Equation 3.47 in which the limits of integration are ± 0.2. Curve (b) is an example of Gibbs oscillation and we observe that, in this case, it has 10 cycles over the frequency span from $-f_v$ to $+f_v$ (Figure 3.12 shows one cycle per 0.1 cycles/data interval) corresponding to $n = 10$, the first term truncated in the Fourier series. Curve (c) is the response function from Equation 3.51, where weights $\overline{w}_k$ are computed using Equation 3.54. The quantity Δf_L (Δf_R) is the bandwidth between f_c and the nearest unit (zero) response. The advantage in using the sigma factor to reduce the Gibbs oscillation is plainly evident. The trade-off with the reduced Gibbs oscillation in curve (c) is the increased width of the transition band, the frequency interval between the nearest unit and zero responses surrounding f_c. This trade-off is such that there is, in general, no advantage to using a sigma factor to a power greater than one.

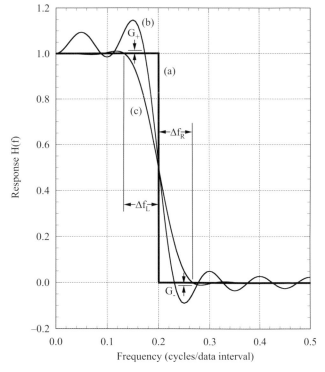

Figure 3.12   (a) Ideal response function. (b) Lanczos filter for $f_c = 0.2$, $2n-1 = 19$ weights, and sigma factor $= 0$. (c) Same as (b) but for sigma factor $= 1$. The Gibbs oscillation associated with (c) is denoted by G_+ and G_-.

CH 3 FILTERING DATA

The approximate properties of the maximum Gibbs oscillation (G_+ or G_- in Figure 3.12) can be summarized by examining three areas in Figure 3.13, in which it should be noted that the vertical axis is the number of weights given by $2n + 1$. We can return to using the traditional formula for total number of weights because if we extend the limits of k from $\pm (n - 1)$ to $\pm n$ in Equation 3.54, the weights $\overline{w}_{\pm n} = 0$. We just need to recognize that the weights at the extrema of the more familiar formula for the total number of weights are zero. To the right of the dashed line, where the number of weights is small or the cut-off frequency is close to the Nyquist frequency, the response function never crosses zero response (i.e., G_- is negative). This is a consequence of $f_v/n + f_c > f_v$. In the large area inside the solid and dashed curved lines the Gibbs oscillation is about 0.01, Δf_R and Δf_L are approximately equal to $1.3/(2n)$, and $H(f_c) \cong 0.5$. The best fidelity is found here. Larger values of Gibbs oscillation occur in an irregular pattern along the border of this area and especially at low cut-off frequencies.

In the small area between the solid and dashed curves on the left side of Figure 3.13 the magnitude of G_+ is zero, since the response function decreases toward zero directly from the origin. Simultaneously, $\Delta f_L = f_c$. The magnitude of G_- can be substantially greater than 0.01 when the number of weights is small. In this region $H(f_c) > 0.5$. Regardless of the area in which the intersection of the cut-off frequency f_c and number of weights $(2n + 1)$ lies, increasing the number of weights always results in a narrowing of the transition band and the associated steepening of the response from unity toward zero.

For high-pass filters, Δf_L (Δf_R) is the bandwidth between f_c and the nearest zero (unit) response. With the convention that G_+ (G_-) is the maximum value of

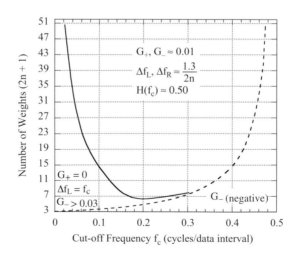

Figure 3.13 The magnitudes of the maximum positive (G_+) and negative (G_-) Gibbs oscillations and the left (Δf_L) and right (Δf_R) bandwidths (Figure 3.12).

the Gibbs oscillation below zero (above unit) response, Figure 3.13 can be applied directly to high-pass filters. In this case, G_ (negative) means that the response function never passes through one. These relations follow from Section 3.2.5, in which a high-pass filter response is one minus the low-pass filter response.

3.4.3 An application

In this section we apply low-pass, high-pass and band-pass filters to hourly temperature data at St. Louis, Missouri, USA, for February 2010. These data are available on the website http://www.wiley.com/go/duchon/timeseriesanalysis. The filename is STL_201002_hrly_temp.xls. See problem 16 in Chapter 1 for information on the data structure. The weight sequences and responses are obtained from a computer program provided in Appendix 3.C for Lanczos filtering in which the required inputs are the cut-off frequency (two frequencies in the case of a band-pass filter), the number of weights, and the type of filter (low-, high-, or band-pass). The power of the sigma factor is set to one but can be easily changed in the program.

The 672 values of temperature are plotted in Figure 3.14. There are five distinct cold front passages with two extended warming trends, one beginning 15 February and the other 25 February. First we examine periods longer than one day. Since one cycle per day $= 0.0417$ cy/h we would like the ideal low-pass response function to drop from one to zero at this frequency. However, we observed from Figures 3.10 and 3.11 that the value of $H(f)$ or $\overline{H}_n(f)$ at the cut-off frequency f_c using the Lanczos filter is about 0.5. Consequently, f_c needs to be adjusted such that it is to the left of one cycle per day in order to achieve zero response at the daily cycle. In addition, because f_c is so close to the origin, we can anticipate using a large number of weights if we wish to reasonably approximate the ideal response. Let us say that we can afford to lose one day at either end of the month. In this case, to stay to the right of the solid line in

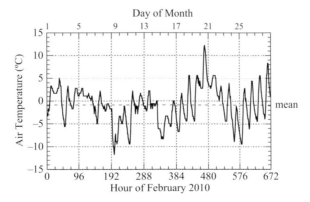

Figure 3.14 Hourly air temperature for February 2010 at Lambert-St. Louis (Missouri) International Airport.

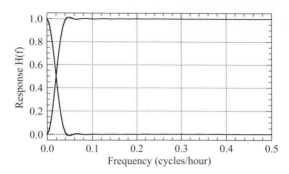

Figure 3.15 Frequency response functions for low-pass and high-pass Lanczos filters for $2n+1=51$ weights and $f_c=0.016$ cycles/hour.

Figure 3.13, we can choose $2n+1=51$ weights. Since the response at f_c is about 0.5 we can estimate f_c from

$$f_c \cong 0.042 - \Delta f_R = 0.042 - 1.3/(2n) = 0.042 - 1.3/50$$
$$= 0.042 - 0.026 = 0.016 \text{ cy/h}. \tag{3.56}$$

The quantity Δf_R (Figure 3.12) in this equation has the approximate value $1.3/2n$ throughout most of the good fidelity region in Figure 3.13. With $2n+1=51$ and $f_c=0.016$ cy/h, the response is essentially zero at 0.042 cy/h, as seen in Figure 3.15 (the low-pass filter response). The low-pass filtered data in Figure 3.16 show mainly the influence of synoptic-scale features: frontal passages and warming and cooling

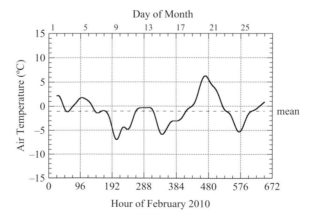

Figure 3.16 The data in Figure 3.14 after applying the filter having the low-pass response $H(f)$ in Figure 3.15. The dashed line is the mean of the filtered data.

trends. Next, we examine periods one day and shorter. The response function for the high-pass filter used here is obtained by subtracting the low-pass filter (Equation 3.51) from one to get

$$\overline{H}'_n(f) = \overline{w}'_0 + 2 \sum_{k=1}^{n} \overline{w}'_k \cos(2\pi f k) \tag{3.57}$$

where $\overline{w}'_0 = 1 - \overline{w}_0$ and $\overline{w}'_k = -\overline{w}_k$ (see Figure 3.15 for high-pass filter response). Figure 3.17 is the result of high-pass filtering. The daily cycle is obvious, in addition to disturbances whose duration is less than a day.

Lastly, consider a band-pass filter. The objective is to obtain a clearer picture of the daily cycle in temperature by eliminating the high frequency fluctuations in Figure 3.17. In order to determine the width of the pass-band it is helpful to examine a variance spectrum of the data. Figure 3.18 shows the periodogram of the St. Louis data out to harmonic 120. The spectrum beyond harmonic 120 is similar to that between about harmonics 100 to 120 with a small average decrease. Judging by the magnitude of the peaks that are a multiple of harmonic 28 (24 hour period), it should be satisfactory to include just those harmonics between 28 and 56 or 0.042 and 0.083 cy/h. The variance at harmonics 84 (eight hour period) and 112 (six hour period) are about two orders of magnitude less than that at harmonic 56 and are comparable to some nearby peaks.

A simple way to create a band-pass filter is to subtract one low-pass filter from another. For example, Figure 3.19a shows the response functions for two low-pass filters with $2n+1 = 43$ weights and $f_{c1} = 0.2$ and $f_{c2} = 0.3$ cycles/data interval;

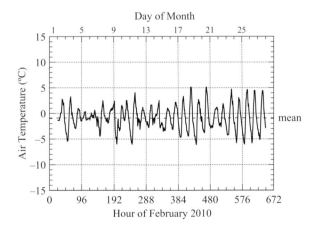

Figure 3.17 The data in Figure 3.14 after applying the filter having the high-pass response H(f) in Figure 3.15 and adding back the mean of the unfiltered data shown by the dashed line.

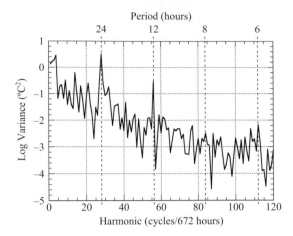

Figure 3.18 Periodogram of the data in Figure 3.14 for harmonics 1-120.

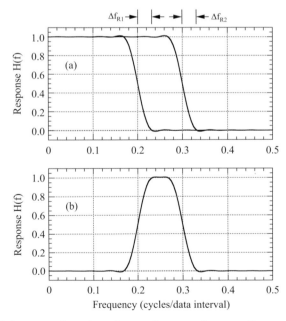

Figure 3.19 (a) Response functions for two low-pass Lanczos filters with $f_c = 0.2$ cy/di, $f_c = 0.3$ cy/di and $2n + 1 = 43$ weights. (b) The band-pass filter that results from the difference of the two response functions in (a).

Figure 3.19b shows the difference. The band-pass smoothed response function is given by

$$\overline{H}_n^*(f) = \overline{w}_0^* + 2 \sum_{k=1}^{n} \overline{w}_k^* \cos(2\pi fk) \qquad (3.58)$$

where $\overline{w}_k^* = \overline{w}_{2k} - \overline{w}_{1k}$, $k = 0, 1, \ldots, n$.

As the width of the pass-band becomes narrower for a given number of weights, the response at the center of the band approaches zero (the two curves in Figure 3.19a move closer together). To keep the response around the center of the band very close to one, the following criterion can be deduced from Figure 3.19a:

$$\Delta f_{R1} + \Delta f_{L2} \leq f_{c2} - f_{c1}. \qquad (3.59)$$

Using the approximation $\Delta f_{R1} = \Delta f_{L2} = 1.3/(2n)$, an equivalent criterion is

$$n \geq 1.3/(f_{c2} - f_{c1}). \qquad (3.60)$$

Thus the narrower the band between cut-off frequencies, the greater n must be to maintain unit response at the band center.

When the objective is to pass essentially a single frequency or a very narrow band of frequencies, Equation 3.59 or Equation 3.60 is especially valuable, in which case the equal sign is used. More generally, we want to pass a band of frequencies with essentially unit response so that the problem then is to select the f_{c1} and f_{c2} that will yield such a response. This is the case for the St. Louis data in which the pass-band is from 0.042 to 0.083 cy/h. Thus,

$$f_{c1} = 0.042 - \Delta f_{R1} \cong 0.042 - 1.3/2n \qquad (3.61a)$$

and

$$f_{c2} = 0.083 + \Delta f_{L2} \cong 0.083 + 1.3/2n. \qquad (3.61b)$$

We are free to choose n insofar as (i) the above relationship between Δf_{R1} and Δf_{L2} and n holds (this means staying away from the frequency origin and the Nyquist frequency when n is small) and (ii) $f_{c1} > 0.0$ and $f_{c2} < f_v$. The greater the number of weights, the narrower the transition bands.

Employing a total of $2n+1 = 71$ weights should provide a good response. From Equation 3.61 the cut-off frequencies are

$$f_{c1} = 0.042 - 1.3/70 = 0.023 \text{ cy/h} \qquad (3.62a)$$

and

$$f_{c2} = 0.083 + 1.3/70 = 0.102 \text{ cy/h}. \qquad (3.62b)$$

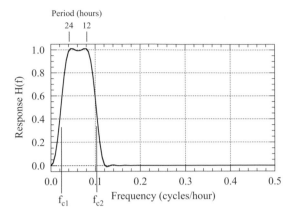

Figure 3.20 Response function for Lanczos band-pass filter with $f_{c1} = 0.023$ cy/di, $f_{c2} = 0.102$ cy/di, and number of weights $2n+1 = 71$.

Using the above values in the computer program in Appendix 3.C produces the band-pass frequency response in Figure 3.20.

The band-pass filtered data are shown in Figure 3.21. The high frequency fluctuations seen in Figure 3.17 have been removed. Even though the variance associated with the semi-daily cycle is an order of magnitude less than that of the daily cycle (Figure 3.18), its impact can be seen in the decreasing portion of the daily cycle for days 18, 19, and 20, for example. The change in shape relative to the nearly straight line (on the scale of the plot) during the increasing portion is brought about by the difference in phase angles between the daily and semi-daily sinusoids.

Problem 7 provides an opportunity to apply Lanczos filtering to the January and July 2009 temperature data at Will Rogers Airport, Oklahoma City, Oklahoma, that were analyzed in problem 16 of Chapter 1.

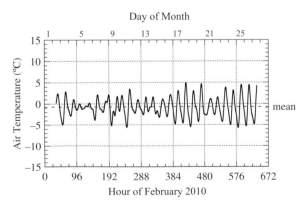

Figure 3.21 The data in Figure 3.14 after applying the bandpass filter having the response H(f) in Figure 3.20 and adding back the mean of the unfiltered data shown by the dashed line.

Appendix 3.A   Convolution of two running mean filters

In this appendix we show how to create a triangular filter by convolving two identical running mean digital filters. The weights for the first running mean filter are given by

$$w_{1m} = \begin{cases} \dfrac{1}{n+1}, & 0 \le m \le n \\ 0, & m > n \end{cases} \tag{3.A.1}$$

and for the second by

$$w_{2m} = \begin{cases} \dfrac{1}{n+1}, & 0 \le m \le n \\ 0, & m > n \end{cases} \tag{3.A.2}$$

for $n > 0$.

Following Equation 3.5 we can write the convolution summation for the weights of the triangular filter as

$$w_k = \sum_{m=-\infty}^{\infty} w_{1m}\, w_{2(k-m)}, \qquad -\infty < k < \infty. \tag{3.A.3}$$

Because we are convolving two weight functions to produce a third weight function, we use the symbol w for all three weight functions.

To demonstrate the convolution, panel (a) below shows two three-weight running mean filters, w_{1m} and w_{2m}, outlined by the solid and dashed lines, respectively, for $n = 2$, with weights $1/(n+1)$. Running mean filter w_{2m} has been reflected about the axis $m = 0$ and is coming from the left and moving to the right to $k = -1$. There is no overlap of $w_{2(-1-m)}$ with w_{1m}.

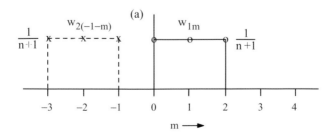

Next, translate $w_{2(-1-m)}$ to the right one time step as shown in panel (b). Now there is overlap.

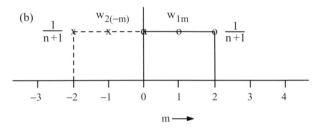

and the weight at $k = 0$ as a consequence of convolution is

$$w_{k=0} = \frac{1}{(n+1)} \times \frac{1}{(n+1)} = \frac{1}{(n+1)^2}. \qquad (3.A.4)$$

Now translate $w_{2(-m)}$ to the right one more time step as shown in panel (c).

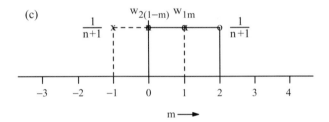

From panel (c) we see that

$$w_{k=1} = \frac{2}{(n+1)^2}. \qquad (3.A.5)$$

If we were to continue with $k = 2$, $k = 3$, and so on, we would conclude the corresponding weights would be

$$w_{k=2} = \frac{3}{(n+1)^2}$$

$$w_{k=3} = \frac{2}{(n+1)^2}$$

$$w_{k=4} = \frac{1}{(n+1)^2}$$

$$w_{k=5} = \frac{0}{(n+1)^2}$$

and so on, for $n = 2$. For the general case of n being a positive integer > 1, we surmise the relation between w_k and n is

$$\vdots$$

$$w_{k=-1} = \frac{0}{(n+1)^2}$$

$$w_{k=0} = \frac{1}{(n+1)^2}$$

$$w_{k=1} = \frac{2}{(n+1)^2}$$

$$\vdots$$

$$w_{k=n-1} = \frac{n}{(n+1)^2}$$

$$w_{k=n} = \frac{n+1}{(n+1)^2}$$

$$w_{k=n+1} = \frac{n}{(n+1)^2}$$

$$\vdots$$

$$w_{k=2n-1} = \frac{2}{(n+1)^2}$$

$$w_{k=2n} = \frac{1}{(n+1)^2}$$

$$w_{k=2n+1} = \frac{0}{(n+1)^2}$$

$$\vdots$$

If we offset the location of the central weight at $k = n$ above to $k = 0$, we obtain a convenient formula for the triangular filter given by

$$w_k = \begin{cases} \dfrac{(n+1) - |k|}{(n+1)^2}, & 0 \leq |k| \leq n \\ 0, & |k| > n \end{cases} \tag{3.A.6}$$

which simplifies to

$$w_k = \begin{cases} \dfrac{1}{(n+1)} \left(1 - \dfrac{|k|}{(n+1)} \right), & 0 \leq |k| \leq n \\ 0, & |k| > n \end{cases} \tag{3.A.7}$$

which is the same as Equation 3.19. The length of the filter is $(2n+1)$ and the sum of the weights is one.

Appendix 3.B   Derivation of Equation 3.20

In this appendix we outline a procedure to derive the response function for a triangular filter. We begin with the weight function for the triangular filter, namely

$$w_k = \begin{cases} \dfrac{1}{n+1}\left(1-\dfrac{|k|}{n+1}\right), & |k| \le n \\ 0, & |k| > n \end{cases} \tag{3.19}$$

The corresponding frequency response function is

$$H(f) = \frac{1}{n+1}\sum_{k=-n}^{n}\left(1-\frac{|k|}{n+1}\right)\exp(-i2\pi fk)$$

$$= \frac{1}{n+1}\sum_{k=-n}^{n}\exp(-i2\pi fk) - \frac{2}{(n+1)^2}\sum_{k=0}^{n}k\cos(2\pi fk) \tag{3.B.1}$$

where $\Delta t = 1$ and the sine term that might otherwise be expected in the final summation is zero since sine is an odd function.

The second summation can be written

$$\sum_{k=0}^{n}k\cos(2\pi fk) = \frac{1}{2\pi}\frac{d}{df}\left(\sum_{k=0}^{n}\sin(2\pi fk)\right)$$

$$= \frac{1}{2\pi}\frac{d}{df}\left(\sum_{k=0}^{n}[\exp(i2\pi fk)-\exp(-i2\pi fk)]/(2i)\right) \tag{3.B.2}$$

$$= \frac{1}{2\pi}\frac{d}{df}\left[\frac{\sin(\pi fn)\sin[\pi f(n+1)]}{\sin(\pi f)}\right]$$

in which Equation 1.B.4 has been used.

Substituting Equation 3.B.2 into Equation 3.B.1 and differentiating yields

$$H(f) = \frac{1}{n+1}\frac{\sin[\pi f(2n+1)]}{\sin(\pi f)} - \frac{1}{(n+1)^2}\left[\frac{(n+1)\sin(\pi fn)\cos[\pi f(n+1)]}{\sin(\pi f)}\right.$$
$$\left. + \frac{n\sin[\pi f(n+1)]\cos(\pi fn)}{\sin(\pi f)} - \frac{\sin(\pi fn)\sin[\pi f(n+1)]\cos(\pi f)}{\sin^2(\pi f)}\right].$$

Expanding the first term on the right to match the form of the second and third terms, then reducing, results in

$$H(f) = \frac{1}{(n+1)^2\sin(\pi f)}\left[\cos(\pi fn)\sin[\pi f(n+1)] + \frac{\sin(\pi fn)\sin[\pi f(n+1)]\cos(\pi f)}{\sin(\pi f)}\right].$$

With additional manipulation, we obtain

$$H(f) = \frac{\sin^2[\pi f(n+1)]}{(n+1)^2\sin^2(\pi f)} \tag{3.B.3}$$

which is the same as Equation 3.20 for $\Delta t = 1$.

Appendix 3.C   Subroutine sigma

```
subroutine sigma (nwt, wt, wtbp, fca, fcb, resp, freq, ihp)
dimension wt(1), resp(1), freq(1), wtbp(1)
data pi, topi /3.1415926536, 6.2831853072/
c******************************************************************
c
c      This subroutine computes the weight sequence and response
c      function for low-pass, high-pass and band-pass Lanczos
c      filters.
c
c
c                              *Input*
c
c....   nwt   Total number of weights, 2n+1. n is the num-
c             ber of weights to the right and left of the central
c             weight. The end weights (n, -n) will always be zero
c             when the sigma factor is greater than 0.
c
c....   fca   The cut-off frequency of the ideal high or low-pass
c             filter.
c
c
```

```
c....   fcb   Used only when a band-pass filter is desired, in which
c             case it is the cut-off frequency for the second low-pass
c             filter. fcb is greater than fca.
c
c....   ihp   If low-pass filter ihp = 0, if high-pass filter ihp = 1,
c             if band-pass filter ihp = 2.
c
c                             *Output*
c
c....   wt    The array of low-, high-, and band-pass computed weights
c             including the central weight and those on either side.
c             Its length is (nwt+1)/2.
c
c....   wt    resp The array of responses at frequency intervals of 0.005
c             cycles/data interval from the origin to the Nyquist
c             frequency. Its length is 101.
c
c....   freq  The array of frequencies at intervals of 0.005 cycles/
c             data interval at which the responses are calculated.
c             Its length is 101.
c
c                             *Other*
c
c....   wtbp  The array of weights for the first low-pass filter
c             used in computing a band-pass filter. Its length is
c             (nwt + 1)/2.
c
c....nsigma The power of the sigma factor. It can be greater than or
c             equal to zero. It is currently set to one.
c
c*****************************************************************
        nsigma = 1
        arg = topi*fca
        argb = topi*fcb
        nw = (nwt - 1) / 2
        anw = nw
        kk = 0
        wt (1) = 2.0*fca
c
c..............Compute weights
c
  91    kk = kk + 1
        do 10 i = 1, nw
        ai = i
        knw = i + 1
```

```
          a = sin(arg*ai) / (pi*ai)
          b = anw*sin(ai*pi / anw) / (pi*ai)
          b = b**nsigma
          wt(knw) = a*b
   10     continue
c
c..............Standardize weights
c
          sum = wt(1)
          do 20 i = 2, knw
   20     sum = sum + 2.0*wt(i)
          do 30 i = 1, knw
   30     wt(i) = wt(i) / sum
          if (kk.ge.2) go to 81
          if (ihp - 1) 1, 2, 3
c
c..............Alter weights to get high-pass filter
c
    2     wt(1) = 1.0 - wt(1)
          do 70 i = 2, knw
   70     wt(i) = -wt(i)
          go to 1
c..............Compute weights of 2nd low-pass filter for band-pass filter
c
    3     do 80 i = 1, knw
   80     wtbp(i) = wt(i)
          arg = argb
          wt(1) = 2.0*fcb
          go to 91
c
c..............Alter weights to get band-pass filter
c
   81     do 90 i = 1, knw
   90     wt(i) = wt(i) - wtbp(i)
c
c..............Compute response function
c
    1     nf = 101
          frqint = 0.5 / float(nf - 1)
          freq(1) = 0.0
          d = 0.0
          do 60 j = 2, knw
   60     d = d + 2.0*wt(j)
          resp(1) = d + wt(1)
          do 40 i = 2, nf
```

```
        ai = i - 1
        freq(i) = ai*frqint
        fr = freq(i)
        d = 0.0
        do 50 j = 2, knw
        bi = j - 1
        d = d + wt(j)*cos(topi*fr*bi)
50      continue
        resp(i) = wt(1) + 2.0*d
40      continue
        return
        end
```

Problems

1 On a graph of response versus frequency, sketch an example of a high-pass filter, a low-pass filter, and a band-stop filter, labeling each. Each curve should extend from 0 to 0.5 cycles per data interval. What is the sum of the weights for each filter?

2 Show, mathematically, that if the mean of a stationary infinite analog time series is to be preserved after applying a filter of finite length, the area of the weight function must be one.

3 Starting with

$$y_t = \sum_{k=-m}^{m} h_k\, x_{t-k}, \quad t = 1, 2, \cdots N$$

where $m \ll N$, show that if h_k is a filter with the sum of the weights equal to zero, then

$$\bar{y} \cong 0$$

in which the overbar represents the mean value of the realization. Explain any approximations you make. Use a sketch to aid you in your analysis.

4 (a) Write down the formula for the weights w_k for a running mean (rectangular) digital filter.

(b) Using the appropriate Fourier transform and Equation 1.B.4, show that the frequency response function for the running mean filter is given by

$$H(f) = \frac{\sin[\pi f(2n + 1)\Delta t]}{(2n + 1)\sin(\pi f \Delta t)}$$

where $\Delta t = 1$.

(c) Now convert H(f) above into a high-pass filter. Sketch the associated high-pass frequency response H′(f) for frequencies from 0 to 0.5 cycles per data interval.

(d) What is the formula for the high-pass weights w_k'? Also, calculate the value of each of the high-pass weights if the total number of weights is five.

5 The following is a sequence of digital data from a time series.

$$y_5 = 2 \qquad y_6 = 3 \qquad y_7 = -1 \qquad y_8 = 0 \qquad y_9 = 1 \qquad y_{10} = 4$$

(a) If the data are to be smoothed with a high-pass filter that is a five-point triangular filter (total number of weights is five), determine the weights and show your method of determination.

(b) Compute the value of the high-pass filtered time series y_t' at all times for which it is feasible.

(c) Plot H(f) for frequencies extending from 0 to 0.5 cycles per data interval for $\Delta t = 1$. Calculate H(f) at a sufficient number of frequencies to clearly define H(f).

(d) On the same graph as in (c), plot the companion low-pass frequency response function using a dashed line.

6 A cosine filter with $(2n+1)$ weights where $n = 3$ is applied to a digital time series. The central weight $w_0 = 1/3$, $w_1 = w_{-1} = 0.75/3$, and $w_3 = w_{-3} = 0$.

(a) What is the value of $w_2 = w_{-2}$ for this low-pass filter?

(b) What is the value of x_t' at all times for which it is calculable given

$$x_6 = 4 \quad x_7 = 5 \quad x_8 = 6 \quad x_9 = 3 \quad x_{10} = 0$$
$$x_{11} = -1 \quad x_{12} = -3 \quad x_{13} = -1 \quad x_{14} = 2$$

7 This problem deals with the application of high-pass and low-pass filters to Oklahoma City, Oklahoma, hourly temperature data for January and July 2009. These data are available on the website http://www.wiley.com/go/duchon/timeseriesanalysis. The filenames are OKC_200901_hrly_temp.xls and OKC_ 200907_hrly_temp.xls. See problem 16 in Chapter 1 for information on the data structure.

(a) Design a low-pass Lanczos filter to pass only periods longer than one day. That is, periods one day and shorter should be removed. Use 51 weights and a sigma factor to power 1. Discuss your design procedure, particularly your selection of f_c, and plot the filter response function to verify that the daily cycle will be removed.

(b) On a single graph with a common temperature scale, plot both the original January 2009 data and the data passed by the filter.

(c) Use a high-pass Lanczos filter to pass only periods 24 hours and shorter. That is, periods 24 hours and shorter should be essentially completely passed. Plot the filter response function.

(d) Plot the data passed by the high-pass filter and the original January 2009 data on one graph. Use a continuous temperature scale for both time series so that the two curves are separate from each other.

(e) What do the low-pass filtered data show? Indicate the times of the cold front passages on the plot of the low-pass filtered data in (b). What is the average number of days between cold front passages (if that's what you see) over the course of the month?

(f) What do the high-pass filtered data show? Comment on the causes of the variable amplitude of the daily cycle. What accounts for the nighttime temperatures often being noisier-looking than the daytime temperatures?

(g) Use the low- and high-pass filters from (a) and (c) to filter the July 2009 data. Create plots of the original and filtered data similar to those you created in (b) and (d).

(h) Compare and contrast the July time series with the January time series, particularly from a meteorological standpoint.

8 Repeat problem 7, but using data from the January and July 2010 files for San Francisco, California. These data are available on the website http://www.wiley.com/go/duchon/timeseriesanalysis. The filenames are SFO_201001_hrly_temp.xls and SFO_201007_hrly_temp.xls. See problem 16 in Chapter 1 for information on the data structure.

References

Duchon, C.E. (1979) Lanczos filtering in one and two dimensions. *J. Applied Meteor.*, **18**, 1016–1022.

Hamming, R.W. (1977) *Digital Filters.* Prentice-Hall, Englewood Cliffs, N. J.

Lanczos, C. (1956) *Applied Analysis.* Prentice-Hall, Englewood Cliffs, N. J.

4

Autocorrelation

One of the goals of a physical scientist is to understand the morphology of natural events. An obvious step that must be taken is to obtain samples in time and space of variables that characterize the physical properties of an event over its lifetime. The fact that an event has a lifetime implies that it evolves in time and/or space, a consequence of which is that successive observations of its properties are related. This is called *autocorrelation*, the term "auto" meaning "with itself." The degree of autocorrelation depends on the physical nature of the phenomenon being sampled and the time and/or space separation between successive observations. Sometimes the term "serial correlation" is used in place of autocorrelation. While both terms have the same meaning, we tend to use the latter term.

An example of a meteorological event is an isolated thunderstorm, the lifetime of which, from birth through maturity to death, may last around an hour and move 50 km. A variable that is used to characterize storm intensity is the maximum speed in the updraft located inside the storm. Initially it is small, reaches a peak at maturity and then decreases. Successive values of the maximum updraft speed are clearly related; that is, they are autocorrelated. Another example is the changing river stage (water level) at a location along a river fed by basin runoff in response to rainfall in the basin. The river stage rises, reaches a maximum, and then falls as the runoff ceases, the record of which is called a hydrograph. Thus the magnitudes of successive measurements of river stage are related.

From a statistical viewpoint, a positive value of autocorrelation for a time separation of 10 minutes, for example, means that a higher-than-average observation tends to be followed by another higher-than-average observation 10 minutes later, and similarly for lower-than-average observations. A negative value means that a higher-than-average observation tends to be followed by a lower-than-average

Time Series Analysis in Meteorology and Climatology: An Introduction, First Edition. Claude Duchon and Robert Hale.
© 2012 John Wiley & Sons, Ltd. Published 2012 by John Wiley & Sons, Ltd.

observation 10 minutes later, and vice versa. The greater the tendency for any two successive departures from the mean to be of the same or opposite sign for a given time separation, the greater the positive or negative autocorrelation, respectively. The sequence of autocorrelation values associated with increasing time separation is called the *autocorrelation function*. While the autocorrelation function is dimensionless, it has a companion function called the *autocovariance function*, the units of which are the squared units of the variable of interest. The autocorrelation function is a standardized autocovariance function.

Any statistical property that involves the number of independent data or degrees of freedom (dof) will be altered by the presence of autocorrelation in the data used to calculate that property. Common examples are the variance, variance of the mean, variance of the variance, and confidence intervals for the population mean, each of which is discussed in this chapter.

Before proceeding further, it may be appropriate to review Section 1.4.1, wherein various statistical concepts and terms are discussed. A number of them will be used throughout this chapter.

4.1 Definition and properties

The autocovariance function (acvf) of a random process denoted by random variable (rv) $X(t)$ is given by

$$\gamma(t_2 - t_1) = \text{Cov}[X(t_1), X(t_2)]$$
$$= E[(X(t_1) - \mu(t_1))(X(t_2) - \mu(t_2))] \tag{4.1}$$

where Cov means covariance, E is the expectation operator, and μ is the time-dependent population mean. In order to understand Equation 4.1 we consider the vertically stacked array of time series shown in Figure 4.1 (which is nearly identical to Figure 1.17). The realizations shown are examples from the population of time series comprising a random process. Taking the expectation in Equation 4.1 means we are finding the average of products of the departures of rv $X(t_1)$ about its population mean $\mu(t_1)$ with those of rv $X(t_2)$ about its population mean $\mu(t_2)$. Each population mean results from taking the average across all members of the population of time series at the respective times and is equivalent to taking the expectation.

Equation 4.1 is the most general form for the population autocovariance function. It allows for nonstationary time series, as reflected by the time dependence of the population means. However, analysis of data with time varying statistical properties can present an enormous challenge. In Section 1.4.1 we discussed the need to transform nonstationary time series to stationary time series by any appropriate method. Accordingly, we continue our investigation of autocorrelation for

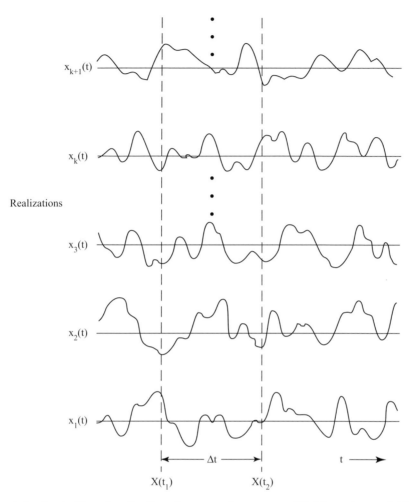

Figure 4.1 A selection of realizations from a random process. $X(t_1)$ is random variable X at time t_1, $X(t_2)$ is random variable X at time t_2. The light horizontal lines are the same reference value of x for each realization.

stationary random processes only. Under this proviso, the expected value of products of the departures from the mean for a given time separation, $\Delta t = t_2 - t_1$ in Figure 4.1, is independent of location along the time axis. That is, the autocorrelation function depends only on time difference, not actual time. As a consequence, we can write the equations for the autocovariance and autocorrelation for a stationary random process directly from Equation 4.1 for both analog and digital time series.

4.1.1 Analog data

For an analog stationary random process X(t), the population autocovariance function given by Equation 4.1 reduces to

$$\gamma(u) = \text{Cov}[X(t), X(t+u)]$$
$$= E[(X(t) - \mu)(X(t+u) - \mu)] \tag{4.2}$$

where u is referred to as the *time lag* or simply *lag*. For a stationary process the population means are no longer dependent on time; they all have the same value. Thus, the expectation in Equation 4.2 depends only on time separation u, not on actual time.

As stated earlier, the population autocorrelation function (acf) is the standardized population autocovariance function, so we can write

$$\rho(u) = \frac{\gamma(u)}{\gamma(0)} = \frac{\gamma(u)}{\sigma_X^2} \tag{4.3}$$

where σ_X^2 is the population variance.

Three common properties of the acf are:

(1) $\rho(0) = 1$

(2) $\rho(u) = \rho(-u)$ (also $\gamma(u) = \gamma(-u)$)

(3) $|\rho(u)| \leq 1$,   for all u.

Property (1) is a consequence of the definition of ρ given by Equation 4.3 and the fact that the autocovariance at lag 0 is identical to the variance, as seen in Equation 4.2. Property (2) follows from the interchangeability of X(t) and X(t + u) in Equation 4.2 because of the stationarity assumption. Thus the acvf and acf are even functions.

Property (3) can be proved by considering the variance of the linear combination $(\alpha_1 Z_1 + \alpha_2 Z_2)$, where α_1 and α_2 are coefficients and Z_1 and Z_2 are random variables. Then, from Equation 1.18,

$$\text{Var}[\alpha_1 Z_1 + \alpha_2 Z_2] = E[\{(\alpha_1 Z_1 + \alpha_2 Z_2) - E[\alpha_1 Z_1 + \alpha_2 Z_2]\}^2]$$
$$= E[\{\alpha_1(Z_1 - \mu_{Z_1}) + \alpha_2(Z_2 - \mu_{Z_2})\}^2]$$
$$= \alpha_1^2 \text{Var}[Z_1] + 2\alpha_1 \alpha_2 \text{Cov}[Z_1, Z_2] + \alpha_2^2 \text{Var}[Z_2]. \tag{4.4}$$

Because Equation 4.4 is positive or zero, dividing both side by α_2^2 yields

$$f(\alpha_1/\alpha_2) = \text{Var}[Z_1](\alpha_1/\alpha_2)^2 + 2\text{Cov}[Z_1, Z_2](\alpha_1/\alpha_2) + \text{Var}[Z_2] \tag{4.5}$$

which is likewise positive or zero. This is a quadratic equation in (α_1/α_2), thus $f(\alpha_1/\alpha_2)$ will form a parabola if plotted on real axes. If $f(\alpha_1/\alpha_2)$ is everywhere positive, then there are no zero crossings and thus no real roots; the two roots of the equation must be complex. This means that the discriminant of Equation 4.5 must be less than zero. If $f(\alpha_1/\alpha2) = 0$, the parabola intersects the (α_1/α_2) axis at a single point (a double root) and the discriminant of Equation 4.5 is equal to zero. Consequently, for Equations 4.4 and 4.5 to always be positive or zero requires the discriminant in Equation 4.5 to be ≤ 0, resulting in

$$\frac{(\mathrm{Cov}[Z_1, Z_2])^2}{\mathrm{Var}[Z_1]\,\mathrm{Var}[Z_2]} \leq 1. \tag{4.6}$$

Replacing Z_1 by $X(t)$ and Z_2 by $X(t+u)$ and taking the square root yields

$$\left|\frac{\mathrm{Cov}[X(t), X(t+u)]}{(\mathrm{Var}[X(t)]\,\mathrm{Var}[X(t+u)])^{1/2}}\right| = \left|\frac{\mathrm{Cov}[X(t), X(t+u)]}{\mathrm{Var}[X(t)]}\right| = \left|\frac{\gamma(u)}{\sigma_X^2}\right| = |\rho(u)| \leq 1$$

which is property (3). This is the approach taken by Jenkins and Watts (1968) to prove property (3). In practice, it is possible to compute sample values of the acf that exceed unity, but this is an artifact of the formulas used for computation and occurs with nonrandom time series, for example, a sinusoid.

4.1.2 Digital data

Expressions parallel to Equation 4.2, Equation 4.3, and the acf properties for digital data follow. The population autocovariance function is

$$\gamma(k) = \mathrm{Cov}[X_t, X_{t+k}]$$
$$= E[(X_t - \mu)\,(X_{t+k} - \mu)], \quad |k| = 0, 1, 2, \ldots \tag{4.7}$$

where k is the lag for unit increments in time. The population autocorrelation function is

$$\rho(k) = \frac{\gamma(k)}{\gamma(0)} = \frac{\gamma(k)}{\sigma_X^2}, \quad |k| = 0, 1, 2, \ldots. \tag{4.8}$$

The three common properties are:

(1) $\rho(0) = 1$

(2) $\rho(k) = \rho(-k)$ (also $\gamma\,(k) = \gamma\,(-k)$)

(3) $|\rho(k)| \leq 1$,   for all k.

4.2 Formulas for the acvf and acf

4.2.1 Acvfs for analog data

When performing statistical analyses in Chapter 1 we used upper case symbols to represent random variables and lower case symbols to represent samples or realizations. We followed this convention in the previous section and will continue to do so throughout this chapter. Thus $c(u)$ and $c'(u)$ below identify working formulas for calculating the acvf of a realization $x(t)$. The structure of Equation 4.2 suggests there are two formulas for analog data. The first is

$$c(u) = \begin{cases} \dfrac{1}{T} \displaystyle\int_0^{T-|u|} (x(t) - \bar{x})(x(t + |u|) - \bar{x})\, dt, & 0 \le |u| < T \\ 0, & |u| \ge T \end{cases} \tag{4.9}$$

and the second is

$$c'(u) = \begin{cases} \dfrac{1}{T - |u|} \displaystyle\int_0^{T-|u|} (x(t) - \bar{x})(x(t + |u|) - \bar{x})\, dt, & 0 \le |u| < T \\ 0, & |u| \ge T \end{cases} \tag{4.10}$$

where

$$\bar{x} = \frac{1}{T} \int_0^T x(t)\, dt.$$

Figure 4.2 demonstrates one way to understand how $c(u)$ or $c'(u)$ is calculated. There is but one time series $x(t)$ extending from 0 to T. It is represented by the upper rectangle in the figure. The identical time series is represented by the lower rectangle, but is shifted u units of time relative to the upper rectangle. Cross-multiplication of the overlapping time series (shaded rectangles) as expressed by the integrals in Equations 4.9 and 4.10 leads to either value of autocovariance $c(u)$ or $c'(u)$. The coefficient $1/(T - |u|)$ of the integral in Equation 4.10 takes into account the

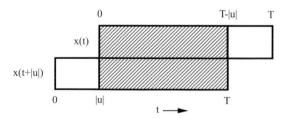

Figure 4.2 Schematic representation of an autocovariance calculation. The shaded area represents the portion of each series involved in the integration of $c'(u)$ or $c(u)$.

continual reduction in the overlap of the two time series as the lag u increases. In Equation 4.9 no accounting is made for the reduction in overlap.

4.2.2 Acvfs for digital data

The form of Equation 4.7 suggests similar formulas for digital data. Expressions parallel to Equations 4.9 and 4.10 are

$$c(k) = \begin{cases} \dfrac{1}{N} \displaystyle\sum_{t=0}^{N-|k|-1} (x_t - \bar{x})(x_{t+|k|} - \bar{x}), & |k| = 0, 1, \ldots, N-1 \\ 0, & |k| > N-1 \end{cases} \qquad (4.11)$$

and

$$c'(k) = \begin{cases} \dfrac{1}{N-|k|} \displaystyle\sum_{t=0}^{N-|k|-1} (x_t - \bar{x})(x_{t+|k|} - \bar{x}), & |k| = 0, 1, \ldots, N-1 \\ 0, & |k| > N-1 \end{cases} \qquad (4.12)$$

where

$$\bar{x} = \frac{1}{N} \sum_{t=0}^{N-1} x_t.$$

Similar to analog data, the coefficient $1/(N - |k|)$ takes into account the continual reduction in the number of products as the lag k increases. In calculating the coefficient in the first formula, no accounting is made for the loss of products. In either Equations 4.11 or 4.12, the acvf is proportional to the sum of the products of a given time series with the same time series lagged k units in time. Figure 4.2 is also a schematic representation of this process if T is replaced by $N-1$, $T-|u|$ by $N-|k|-1$, $x(t)$ by x_t, $x(t+|u|)$ by $x_{t+|k|}$, and $|u|$ by $|k|$.

4.2.3 Mean square error of acvf estimators

As given above, $c(k)$ and $c'(k)$ are formulas to use with realizations of digital data. If we now consider $c(k)$ and $c'(k)$ to be estimators (which are random variables) of the population acvf γ, then x_t, $x_{t+|k|}$ and $\bar{x}$ must be treated as random variables. By our convention, upper case notation is used for random variables. Hence, sample $c(k)$ becomes random variable $C(k)$, sample $c'(k)$ becomes random variable $C'(k)$, sample x_t becomes random variable X_t, and sample $\bar{x}$ becomes random variable $\bar{X}$.

A logical question to ask is which estimator is better to use. The answer can be approached by comparing their mean square errors. The general expression for the mean square error of an estimator θ of parameter Θ is given by

$$
\begin{aligned}
E[(\theta - \Theta)^2] &= E[\{(\theta - E[\theta]) + (E[\theta] - \Theta)\}^2] \\
&= Var[\theta] + B^2(\theta)
\end{aligned}
\tag{4.13}
$$

where B represents bias. Thus, the mean square error of an estimator is the sum of the variance of the estimator about its expected value and the square of the bias. The cross-product term is zero.

If we treat Equations 4.11 and 4.12 as estimators for the acvf and apply Equation 4.13 to each, where $C(k)$ or $C'(k)$ corresponds to θ and $\gamma(k)$, the autocovariance function for a stationary random process, corresponds to Θ, the results are

$$
E[(C(k) - \gamma(k))^2] = Var[C(k)] + B^2[C(k)]
\tag{4.14}
$$

and

$$
E[(C'(k) - \gamma(k))^2] = Var[C'(k)] + B^2[C'(k)].
\tag{4.15}
$$

We first determine the bias of Equation 4.11 by taking its expectation

$$
\begin{aligned}
E[C(k)] &= E\left[\frac{1}{N}\sum_{t=0}^{N-|k|-1}(X_t - \overline{X})(X_{t+|k|} - \overline{X})\right], \quad |k| = 0, 1, 2, \ldots, N-1 \\
&= E\left[\frac{1}{N}\sum_{t=0}^{N-|k|-1}((X_t - \mu) - (\overline{X} - \mu))((X_{t+|k|} - \mu) - (\overline{X} - \mu))\right]
\end{aligned}
\tag{4.16}
$$

in which the population mean μ, has been introduced. Expanding the summation and taking the expectation yields

$$
\begin{aligned}
E[C(k)] = \left(1 - \frac{|k|}{N}\right)&\left\{\gamma(k) + Var[\overline{X}]\right. \\
&- E\left[(\overline{X} - \mu)\frac{1}{N - |k|}\sum_{t=0}^{N-|k|-1}(X_{t+|k|} - \mu)\right] \\
&\left.- E\left[(\overline{X} - \mu)\frac{1}{N - |k|}\sum_{t=0}^{N-|k|-1}(X_t - \mu)\right]\right\}.
\end{aligned}
\tag{4.17}
$$

The comparable expression for the other choice of estimator is

$$E[C'(k)] = \gamma(k) + \text{Var}[\overline{X}]$$

$$- E\left[(\overline{X} - \mu)\frac{1}{N - |k|} \sum_{t=0}^{N-|k|-1} (X_{t+|k|} - \mu)\right]$$

$$- E\left[(\overline{X} - \mu)\frac{1}{N - |k|} \sum_{t=0}^{N-|k|-1} (X_t - \mu)\right]. \qquad (4.18)$$

The two expectations on the right-hand sides of Equations and 4.18 are equal to each other and when $k = 0$ their sum is $-2\,\text{Var}[\overline{X}]$ in each equation. Thus, for $k = 0$ and removal of the curly brackets in Equation , the sum of the last three terms on the right-hand sides of Equations and 4.18 is $-\text{Var}[\overline{X}]$. As $|k|$ increases, the sum of the same terms will become less negative but will be always less than $\text{Var}[\overline{X}]$. Therefore, in general, both $C'(k)$ and $C(k)$ are biased. However, if N is sufficiently large, such that $\overline{X} \approx \mu$, the sums of the three terms will be small, resulting in

$$E[C(k)] \approx \left(1 - \frac{|k|}{N}\right)\gamma(k) \qquad (4.19)$$

and

$$E[C'(k)] \approx \gamma(k). \qquad (4.20)$$

The corresponding bias squared terms are

$$B^2[C(k)] \approx \left(\frac{|k|}{N}\gamma(k)\right)^2 \qquad (4.21)$$

and

$$B^2[C'(k)] \approx 0. \qquad (4.22)$$

As seen in Equation 4.19, the effect of bias is to systematically reduce the magnitude of the expected acvf relative to the process acvf as $|k|$ increases. There is no dependence of bias on lag in Equation 4.20. With the assumption of N sufficiently large, $C(k)$ and $C'(k)$ are referred to as biased and unbiased estimators, respectively.

At this point we've determined the bias squared portion of the mean square error. Because the determination of the variance of each acvf estimator is complex, its development is postponed until Section 4.7. In summary of that section, it is shown that the variance ratio $\text{Var}[C'(k)]/\text{Var}[C(k)]$ is approximately $N^2/(N - |k|)^2$ so that the variability of the $C'(k)$ estimator relative to that of $C(k)$ becomes increasingly unstable as lag k increases. It is thought that for most acvfs the effect of the increasing variance of $C'(k)$ overwhelms the bias squared effect associated with $C(k)$ (Jenkins and Watts, 1968, pp. 179–180). Consequently, we recommend using Equations 4.9 or 4.11. Also, we observe that in estimating acvfs at small values of lag, numerical differences between the two pairs of formulas will be small.

4.2.4 Acfs for analog and digital data

In parallel with Equations 4.9 and 4.10, the sample acfs for analog data are

$$r(u) = \frac{c(u)}{c(0)} = \frac{c(u)}{s^2}, \qquad |u| \leq T \tag{4.23}$$

$$r'(u) = \frac{c'(u)}{c'(0)} = \frac{c'(u)}{s^2}, \qquad |u| \leq T \tag{4.24}$$

where the sample variance is given by

$$s^2 = \frac{1}{T} \int_0^T (x(t) - \bar{x})^2 \, dt.$$

For digital data the sample acfs are

$$r(k) = \frac{c(k)}{c(0)} = \frac{c(k)}{s^2}, \qquad |k| = 0, 1, 2, \cdots \tag{4.25}$$

and

$$r'(k) = \frac{c'(k)}{c'(0)} = \frac{c'(k)}{s^2}, \qquad |k| = 0, 1, 2, \cdots \tag{4.26}$$

where the sample variance is given by

$$s^2 = \frac{1}{N} \sum_{i=0}^{N-1} (x_i - \bar{x})^2.$$

4.3 The acvf and acf for stationary digital processes

In Chapter 2, which dealt mainly with analog signals, we saw that a linear time-invariant system consists of an input signal that is modified to produce an output signal as expressed through the convolution integral

$$y(t) = \int_{-\infty}^{\infty} h(u) x(t - u) \, du. \tag{2.1}$$

Figure 4.3 Input digital stationary process Z_t passing through system function h_t to yield output process X_t.

That the system is stable requires that the integral of the absolute system function $|h(u)|$ be finite (Section 2.1). Now we replace the analog input signal $x(t)$ in Equation 2.1 by a *stationary digital process*, Z_t, to yield an output process, X_t, as shown in Figure 4.3. We will use this notation in both this chapter and Chapter 5. In parallel with analog data, the equivalent criterion for the output process for digital data to be stationary again requires that the sum of the sequence of absolute weights be finite; that is,

$$\sum_{t=-\infty}^{\infty} |h_t| \leq K < \infty$$

where K is a constant. For physical realizability, that is, the current output value cannot depend on future input values, the expression for the output process is, by analogy with Equation 2.1,

$$X_t = \sum_{i=0}^{\infty} h_i \, Z_{t-i}$$

$$= h_0 \, Z_t + h_1 \, Z_{t-1} + \dots. \tag{4.27}$$

Taking the expectation of Equation 4.27 yields

$$E[X_t] = \mu_X = \sum_{i=0}^{\infty} h_i \, E[Z_{t-i}] = \mu_Z \sum_{i=0}^{\infty} h_i$$

which shows that the mean of the output process is equal to the product of the mean of the input process and the sum of the system function weights.

By substituting Equation 4.27 into Equation 4.2 we obtain the relationship between the acvf of the output process and the acvf of the input process

$$\gamma_X(k) = E[(X_t - \mu_X)(X_{t+k} - \mu_X)]$$

$$= E\left[\sum_{i=0}^{\infty} h_i \, (Z_{t-i} - \mu_Z) \sum_{j=0}^{\infty} h_j \, (Z_{t+k-j} - \mu_Z)\right]$$

$$= \sum_{i=0}^{\infty} \sum_{j=0}^{\infty} h_i \, h_j \, E[(Z_{t-i} - \mu_Z)(Z_{t+k-j} - \mu_Z)]$$

$$\gamma_X(k) = \sum_{i=0}^{\infty} \sum_{j=0}^{\infty} h_i \, h_j \, \gamma_Z(k - j + i). \tag{4.28}$$

In the special case when Z_t is white noise, that is, $E[Z_t] = 0$ and $Var[Z_t] = E[Z_t^2] = \sigma_Z^2$, Equation 4.27 becomes the expression for the *general linear process*. It is a stationary random process defined by

$$X_t - \mu_X = \sum_{i=0}^{\infty} h_i \, Z_{t-i} \tag{4.29}$$

in which Z_t is white noise. It is referred to as a "general linear" process because it comprises a linear combination of, potentially, an infinite number of white noise random variables multiplied by appropriate weights to create X_t with virtually any valid statistical structure. Note that the left-hand side now represents departures from the population mean μ_X, the value of which is arbitrary. This is in contrast to Equation 4.27, wherein the output process mean is directly related to the input process mean. The form of Equation 4.29 simply allows the output process to have any desired mean, with the departures from this mean determined by the zero-mean white noise input. The acvf for white noise process Z_t, as it would be applied in Equation 4.28, is

$$\gamma_Z(k - j + i) = \begin{cases} \sigma_Z^2, & k - j + i = 0 \\ 0, & k - j + i \neq 0 \end{cases} \tag{4.30}$$

since the autocorrelation at any nonzero lag is zero for a white noise process. The acvf for the general linear process is then

$$\gamma_X(k) = \sigma_Z^2 \sum_{i=0}^{\infty} h_i \, h_{i+|k|}, \quad |k| = 0, 1, 2, \ldots . \tag{4.31}$$

Equation 4.31 can be verified by substituting values of k into Equation 4.28. Similarly, the acf is

$$\rho_X(k) = \frac{\gamma_X(k)}{\gamma_X(0)} = \frac{\sum\limits_{i=0}^{\infty} h_i \, h_{i+|k|}}{\sum\limits_{i=0}^{\infty} h_i^2}, \quad |k| = 0, 1, 2, \ldots . \tag{4.32}$$

What we have done in this section is to expand the input–output relationship for linear systems developed in Chapter 2 to input–output stationary random processes. Equation 4.28 shows the relationship between the output and input acvfs. When the input process is restricted to white noise, the resulting output process is called the general linear process. The output acvf in Equation 4.31 is the product of the white noise variance and the sum of the products of the weights. The data model in Equation 4.29 has many practical applications in science and econometrics. In the next section we derive the acvfs and acfs for two data models and apply a white noise test to each acf.

4.4 The acvf and acf for selected processes

4.4.1 White noise

The simplest model for data is digital white noise. In Equation 4.29 let $h_0 = 1$ and $h_i = 0$ for $i \neq 0$ so that

$$X_t - \mu = Z_t \tag{4.33}$$

where μ is understood to be the population mean of X_t (previously μ_x), since the population mean of Z_t is zero. From Equation 4.31,

$$\gamma_X(k) \equiv E[Z_T \, Z_{t+k}] = \begin{cases} \sigma_Z^2 = \sigma_X^2, & k = 0 \\ 0, & k \neq 0 \end{cases}$$

resulting in the acf

$$\rho_X(k) = \frac{\gamma_X(k)}{\sigma_X^2} = \begin{cases} 1, & k = 0 \\ 0, & k \neq 0 \end{cases}. \tag{4.34}$$

The population acf in Equation 4.34 is plotted in Figure 4.4 along with the acf (using Equation 4.25) from a realization of 100 white noise values (which will be discussed in Section 4.4.4).

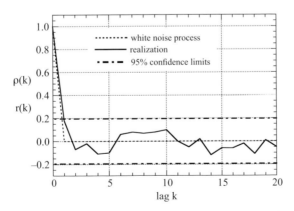

Figure 4.4 The acf $\rho(k)$ for a white noise process (dashed line) and the observed acf $r(k)$ from a realization of 100 values of white noise for $k \leq 20$ (solid line). The 95% confidence limits are also shown.

4.4.2 First-order autoregression

The second model selected for discussion is a first-order autoregressive process, denoted AR(1), given by

$$X_t - \mu = \alpha(X_{t-1} - \mu) + Z_t \qquad (4.35)$$

where Z_t is white noise as defined above and α is a constant such that $|\alpha| < 1$. The current value, $X_t - \mu$, is some proportion of the previous value, $X_{t-1} - \mu$, plus a random component. It is through α that the process and, hence, a realization of it exhibit autocorrelation. Furthermore, α is similar to the slope coefficient in conventional simple linear regression $y = mx + b$, where α corresponds to m. The analogy to simple linear regression is complete when $X_t - \mu$ replaces y, $X_{t-1} - \mu$ replaces x, and Z_t replaces b, except that rv X_t is regressed onto itself at the previous time step; thus the prefix "auto." The random process defined by Equation 4.35 is equivalent to a linear filter (see Equation 3.3) and has been found to be useful as a data model in many fields of science because of its simple autocorrelation structure and the capability to control the magnitude of autocorrelation with parameter α.

To find the acvf and acf of Equation 4.35 we first show that the AR(1) is a linear process so that we can use Equations 4.31 and 4.32. By recursively solving for $(X_{t-1} - \mu)$, $(X_{t-2} - \mu)$, $(X_{t-3} - \mu)$, and so on, as shown below with $\mu = 0$ for convenience, Equation 4.35 acquires the form of Equation 4.29. Since Equation 4.29 is a linear process, so also is Equation 4.35.

$$X_t = \alpha X_{t-1} + Z_t$$
$$X_t = \alpha(\alpha X_{t-2} + Z_{t-1}) + Z_t$$
$$X_t = \alpha(\alpha(\alpha X_{t-3} + Z_{t-2}) + Z_{t-1}) + Z_t$$

$$\vdots$$

$$X_t = \alpha^0 Z_t + \alpha^1 Z_{t-1} + \alpha^2 Z_{t-2} + \dots.$$

The recursion can be expressed by Equation 4.29 where $h_i = \alpha^i$ $(i \geq 0)$. Thus, from Equation 4.31,

$$\gamma_X(k) = \sigma_Z^2 \sum_{i=0}^{\infty} \alpha^i \, \alpha^{i+|k|}$$

$$= \sigma_Z^2 \, \alpha^{|k|} \sum_{i=0}^{\infty} \alpha^{2i}. \qquad (4.36)$$

It should be observed that, for stationarity

$$\sum_{i=0}^{\infty} |\alpha|^i$$

must be a convergent series, thus $|\alpha| < 1$.

Using Equation 4.31, an alternative form of Equation 4.36 is

$$\gamma_X(k) = \gamma_X(0)\, \alpha^{|k|} = \sigma_X^2\, \alpha^{|k|} \tag{4.37}$$

so that using Equation 4.32, the acf for an AR(1) process becomes

$$\rho_X(k) = \alpha^{|k|} \tag{4.38}$$

and, therefore

$$\alpha = \rho_X(1). \tag{4.39}$$

Furthermore, it follows from equating Equation 4.36 and Equation 4.37 and summing an infinite geometric series that

$$\sigma_X^2 = \frac{\sigma_Z^2}{(1 - \alpha^2)} = \frac{\sigma_Z^2}{(1 - \rho_X^2(1))}. \tag{4.40}$$

We see immediately that as $|\alpha|$ approaches one, the variance of the AR(1) process will become very large.

The acf of the X_t process with $\alpha = 0.75$ is plotted in Figure 4.5 along with the acf from a realization of 100 values. As in Figure 4.4, the actual values in each curve are connected with straight-line segments. Note that in Figure 4.5 the sample acf shows an oscillatory structure. This is characteristic of realizations of acfs when the parent acf (that of the random process) does not rapidly damp out to zero, and occurs because the sample acf is itself serially correlated in the lag domain.

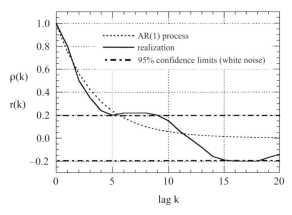

Figure 4.5   The acf $\rho(k)$ for an AR(1) process with $\alpha = 0.75$ (dashed line) and the observed acf r(k) for a realization of 100 values from the same process (solid line). Only acf values for lags $\leq$ 20 are shown. The 95% confidence limits for white noise are also shown.

An alternative way to get Equations 4.37 and 4.38 is to multiply both sides of Equation 4.35 by $(X_{t-k}-\mu)$ and take the expectation, using Equation 4.7 as a guide. Thus,

$$\gamma_X(k) = E[(X_t - \mu)\,(X_{t-k} - \mu)] = E[\alpha(X_{t-1} - \mu)\,(X_{t-k} - \mu) + Z_t(X_{t-k} - \mu)], \quad k = 1, 2, \ldots.$$

Since the current value of white noise rv Z_t is uncorrelated with past values of the output random variables X_{t-k}, $k = 1, 2, \ldots,$

$$
\begin{aligned}
\gamma_X(1) &= \alpha\gamma_X(0)\\
\gamma_X(2) &= \alpha\gamma_X(1) = \alpha^2\gamma_X(0)\\
\gamma_X(3) &= \alpha\gamma_X(2) = \alpha^3\gamma_X(0)\\
&\vdots\\
\gamma_X(k) &= \alpha^{|k|}\gamma_X(0), \qquad |k| = 0, 1, 2, \ldots,
\end{aligned}
\tag{4.41}
$$

from which the population acf is

$$
\begin{aligned}
\rho(1) &= \alpha\\
\rho(2) &= \alpha^2\\
\rho(3) &= \alpha^3\\
&\vdots\\
\rho(k) &= \alpha^{|k|}, \qquad |k| = 0, 1, 2, \ldots.
\end{aligned}
\tag{4.42}
$$

An AR(1) process with positive α is sometimes called a *red noise* process in comparison to a white noise process. The reason is that, as Figure 4.5 shows, the correlation is higher at low-numbered lags than at high-numbered lags. The resulting slowly varying fluctuations mean that the periodogram of an AR(1) process will show greater variance at low frequencies or long wavelengths, analogous to red light occurring at the low frequency or long wavelength end of the optical spectrum. In comparison, there is no preferred structure in the acf for white noise (Figure 4.4) for $|k| > 0$, so that its periodogram will show no preference for variance at any frequency in analogy to white light showing no color preference.

An AR(1) process includes the possibility of α having a negative value, although it is difficult to imagine a physical process that has a negative lag 1 autocorrelation. If such a process exists, we can see from Equation 4.42 that its autocorrelation function would exhibit alternating positive values at even lags and negative values at odd lags with their magnitudes decreasing toward zero with increasing lag. A similar regime of alternating positive and negative values will tend to occur in an observed time series as well.

4.4.3 Second-order autoregression

A second-order autoregression has two regression coefficients. The first coefficient is associated with the output variable one time-step back, as in an AR(1), and the second coefficient is associated with the output variable two time-steps back. Thus the equation of an AR(2) process is

$$X_t - \mu = \alpha_1(X_{t-1} - \mu) + \alpha_2(X_{t-2} - \mu) + Z_t. \tag{4.43}$$

An AR(2) is useful for modeling phenomena that exhibit quasi-periodic behavior. While the autoregressive model can be of any order (third-order has three regression coefficients, fourth-order has four, etc.), only the AR(1) and AR(2) appear to be useful models for physical processes. Problems 7 and 9 at the end of this chapter ask you to derive the variance for an AR(2) process, generate a realization, and compute both the process acf and sample acf.

4.4.4 White noise test on an acf

The objective is to test the null hypothesis that a data set, as viewed through the acf, comes from a population of white noise. This test is complementary to the white noise test developed in Chapter 1 that was applied to periodogram variances. If the null hypothesis is rejected, there is significant autocorrelation in the data set. If the hypothesis cannot be rejected, the data set can be considered to be a realization from a white noise or "purely" random process.

It has been shown by Anderson (1942) that for a normal white noise process the estimator of the autocorrelation function $R(k) = C(k)/C(0)$ has a normal distribution with variance

$$\text{Var}\,[R(k)] \approx \frac{1}{N}, \qquad 0 < |k| \le m \ll N \tag{4.44}$$

where N, the number of data in the realization, is moderate to large. Note, further, that this relation applies only for lags $m \ll N$.

The short dashed lines in Figure 4.6 are ± 1 and ± 1.96 standard deviations in which $R(k)$, representing the estimator of the population acf, has been replaced by $r(k)$, the acf of a realization, since the latter would be plotted in practice. The wider pair of limits corresponds to the *a priori* 95% confidence interval. The interpretation is that there is only one chance in 20 that any *randomly selected* $r(k)$ will lie outside this interval under the white noise null hypothesis.

When Equation 4.44 is applied to Figures 4.4 and 4.5, in which the 95% confidence limits for sample size $N = 100$ are $\pm 1.96/\sqrt{N} = \pm 0.196$, the null hypothesis that the realization in Figure 4.4 comes from a Gaussian white noise hypothesis cannot be rejected, but the same hypothesis applied to Figure 4.5 is easily rejected, as expected.

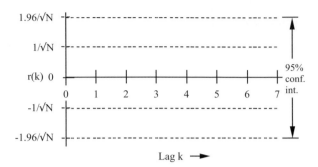

Figure 4.6 Schematic representation of the approximate *a priori* 95% confidence interval for sample autocorrelation coefficients for white noise.

Even though only *a priori* limits are shown, the latter conclusion is reasonable because 12 of the autocorrelation values lie outside the 95% confidence interval, many more than expected by chance, and there is a systematic variation of autocorrelation with lag.

In general, rejection or nonrejection of the white noise null hypothesis using only the *a priori* confidence limits may not be as obvious when real data are tested. Consequently, it is appropriate to develop *a posteriori* confidence limits (Section 1.4.6) for a white noise test applied to a correlation function. A practical *a posteriori* test can be derived from the statistical property that a χ^2 variable with m dof can be created from the sum of the squares of m standard normal variables (Appendix 1.C).

Consider the χ^2 variable with one dof

$$\chi_1^2 = \frac{(R(k) - 0)^2}{\text{Var}\,[R(k)]} \approx NR^2(k) \tag{4.45}$$

in which Equation 4.44 is the source of the standard normal variable. If we sum over m similar chi-square variables, then, from Equation 1.C.11,

$$\chi_m^2 = N\sum_{k=1}^{m} R^2(k). \tag{4.46}$$

Based on Equation 4.46, Box and Jenkins (1970, p. 291) proposed a test statistic

$$Q = N\sum_{k=1}^{m} r^2(k) \tag{4.47}$$

to test for randomness.

As illustrations of using the Q-statistic, consider the computer-generated white noise and AR(1) realizations in Figures 4.4 and 4.5, respectively, each of which have $N = 100$ values. Using the first $m = 20$ values of $r(k)$ we find

$$Q = 13 \text{ for white noise}$$

and

$$Q = 155 \text{ for AR(1)}.$$

Employing a one-tail test (because the concern is not with observing small values of r, but large values) with $\alpha = 5\%$ results in nonrejection of the white noise hypothesis applied to Figure 4.4 and rejection when applied to Figure 4.5, where $\chi^2_{19}(1 - \alpha) = 30.1$ The reason for using 19 dof is that the sample mean was used in estimating $R(k)$. In general, whether one uses m or m − 1 dof will have little consequence if m is at least as large as 20. Ljung and Box (1978) discuss a modified version of the Q-statistic that provides a better test for cases of small sample size.

4.5 Statistical formulas

In introductory texts in statistics, the formula we usually see for the variance of the mean of rv X is

$$\sigma_{\bar{X}}^2 = \frac{\sigma_X^2}{N}.$$

In this formula N is the number of independent data or degrees of freedom (dof) used in calculating the mean. The greater the sample size, the smaller the variance of the mean relative to the variance of the population random variable. Also, introductory texts in statistics usually do not analyze variables that are ordered in time or space, a situation for which the assumption of independent data is often unrealistic. We know from experience that when physical data are collected in time or space, they are typically serially correlated and, therefore, not independent. Dependence or correlation in a time series can substantially increase the variance of the mean relative to assuming independent data. The goal of the next section is to derive the formula for the variance of the estimator for the sample mean when the data are serially correlated. The formula will show us the connection between the number of independent data or degrees of freedom in a realization and the amount of serial correlation present. Then we will determine formulas for unbiased estimators of the mean and variance of a realization in relation to its serial correlation.

4.5.1 Variance of the mean

Consider the running mean process in which

$$\overline{X}_t = \sum_{i=0}^{N-1} \frac{1}{N} X_{t-i}. \tag{4.48}$$

We can link Equation 4.48 to Equation 4.27 by replacing rv X_t with rv $\overline{X}_t$ and rv Z_{t-i} with rv X_{t-i}, where

$$h_i = \begin{cases} 0, & i < 0 \\ \dfrac{1}{N}, & 0 \le i \le N-1 \\ 0, & i > N-1 \end{cases}$$

and apply Equation 4.28 to obtain

$$\gamma_{\overline{X}}(k) = \sum_{i=0}^{\infty} \sum_{j=0}^{\infty} h_i\, h_j\, \gamma_X(k - j + i). \tag{4.49}$$

For $k = 0$ (and a stationary process),

$$\gamma_{\overline{X}}(0) = \mathrm{Var}[\overline{X}_t] = \mathrm{Var}[\overline{X}]$$

$$= \frac{1}{N^2} \sum_{i=0}^{N-1} \sum_{j=0}^{N-1} \gamma_X(i - j). \tag{4.50}$$

If we consider a plane delineated by orthogonal axes i and j, the double summation is over the area outlined in Figure 4.7 by the limits of the summations. Since γ_X is constant along any diagonal $(i - j = m)$, the double sum reduces to the single sum

$$\mathrm{Var}[\overline{X}] = \frac{1}{N^2} \sum_{m=-(N-1)}^{N-1} (N - |m|)\, \gamma_X(m)$$

$$= \frac{1}{N} \left[\gamma_X(0) + \frac{2}{N} \sum_{m=1}^{N-1} (N - m)\, \gamma_X(m) \right]. \tag{4.51}$$

Now that we've lost connection to the k notation in Equation 4.49 in developing the above relation, we can revert back to its previous use of being identified with lag number. Thus,

$$\mathrm{Var}[\overline{X}] = \frac{\sigma_X^2}{N} \left[1 + \frac{2}{N} \sum_{k=1}^{N-1} (N - k)\, \rho_X(k) \right]. \tag{4.52}$$

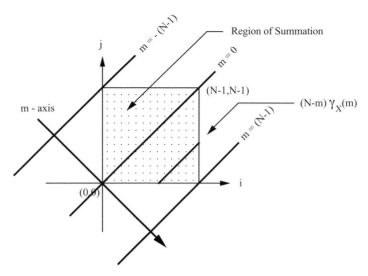

Figure 4.7 Region covered by the double summation in Equation 4.50.

We see that the variance of the mean is given by the product of the variance of the mean for independent data and a scale factor that accounts for their lack of independence. The scale factor is one when the autocorrelation function is everywhere zero for all lags greater than zero. We can preserve the form of the traditional expression for the variance of the mean by introducing a new term N_e that we will call the *effective* or *equivalent* dof and rewrite Equation 4.52 as

$$\sigma_{\bar{X}}^2 = \frac{\sigma_X^2}{N_e} \tag{4.53}$$

where

$$N_e = N\left[1 + \frac{2}{N}\sum_{k=1}^{N-1}(N-k)\rho_X(k)\right]^{-1}. \tag{4.54}$$

The divisor in Equation 4.54 is always greater than or equal to one so that $N_e \leq N$. The greater the autocorrelation in a realization, the smaller N_e is relative to N and the fewer the number of independent data. Fewer independent data (equivalent dof) result in a wider distribution of mean values among realizations.

Equation 4.54 shows the link between number of data, N, in a realization and the equivalent dof, N_e. The link is the autocorrelation function $\rho_X(k)$ of the random process. Unfortunately, it is never known except in modeling studies or simulations. Therefore, in practice, one option is to approximate $\rho_X(k)$ with $r_x(k)$ as determined from a realization. Another option is to fit a smooth curve, say $r_x^*(k)$, to the observed $r_x(k)$ with the constraint that $r_x^*(0) = 1$. If physical considerations provide some insight as to an appropriate $r_x^*(k)$, so much the better.

4.5.2 Mean and variance

To refresh our understanding of the population mean and population variance, we can return to Figure 4.1. When we take the expectation of random variable X(t) for a stationary process, we get $E[X(t)] = \mu$, which is obtained by averaging vertically at any time t across all members of the population of time series. For stationary data, μ is independent of time. Similarly, if we want the population variance, we average the square of the departures from μ, again, vertically across all members of the population; that is, $Var[X(t)] = E[(X(t) - \mu)^2]$. Although Figure 4.1 strictly applies to analog data, the discussion above applies equally well to digital data.

In practice, however, we are confronted with estimating these population measures from a single, or perhaps a few, time series. Nevertheless, we would like to develop a formula for estimating the population mean and population variance using estimators that, in the expected sense, will be unbiased in the presence of autocorrelated data. This is where we begin.

For autocorrelated data, the estimator for the population mean is the same as for uncorrelated data. That is,

$$\overline{X} = \frac{1}{N}\sum_{i=1}^{N} X_i \qquad\qquad (4.55)$$

where N is the number of data. Taking the expectation of Equation 4.55 yields $E[\overline{X}] = E[X_i] = \mu$ and, consequently, the usual estimator for the mean is unbiased.

Consider the following estimator for the population variance:

$$S^2 = \frac{1}{N}\sum_{i=1}^{N} (X_i - \overline{X})^2. \qquad\qquad (4.56)$$

To determine if Equation 4.56 is biased, we take its expected value

$$E[S^2] = \frac{1}{N} E\left[\sum_{i=1}^{N} (X_i - \overline{X})^2\right]. \qquad\qquad (4.57)$$

The summation can be written

$$\sum_{i=1}^{N}(X_i - \overline{X})^2 = \sum_{i=1}^{N}(X_i - \mu - \overline{X} + \mu)^2$$

$$= \sum_{i=1}^{N}(X_i - \mu)^2 - 2(\overline{X} - \mu)\sum_{i=1}^{N}(X_i - \mu) + \sum_{i=1}^{N}(\overline{X} - \mu)^2$$

$$= \sum_{i=1}^{N}(X_i - \mu)^2 - 2N(\overline{X} - \mu)(\overline{X} - \mu) + N(\overline{X} - \mu)^2$$

which reduces to

$$\sum_{i=1}^{N} (X_i - \overline{X})^2 = \sum_{i=1}^{N} (X_i - \mu)^2 - N(\overline{X} - \mu)^2. \qquad (4.58)$$

The expected values of the terms on the right side of Equation 4.58 are $N\sigma_X^2$ and $N\sigma_{\overline{X}}^2 = N\sigma_X^2/N_e$, respectively. Thus, Equation 4.57 becomes

$$E[S^2] = \frac{1}{N} \left(N\sigma_X^2 - \frac{N}{N_e} \sigma_X^2 \right)$$

$$= \sigma_X^2 \frac{N_e - 1}{N_e}. \qquad (4.59)$$

Therefore, Equation 4.56 is a biased estimator. However, from Equation 4.59 we can rewrite estimator Equation 4.56 in an unbiased form as

$$S^2 = \frac{1}{N} \left(\frac{N_e}{N_e - 1} \right) \sum_{i=1}^{N} (X_i - \overline{X})^2. \qquad (4.60)$$

This shows that Equation 4.56 is biased on two accounts. In the first place, even if the data were uncorrelated ($N_e = N$), the unbiased estimator would be

$$S^2 = \frac{1}{N - 1} \sum_{i=1}^{N} (X_i - \overline{X})^2. \qquad (4.61)$$

The effect of the coefficient $1/(N-1)$ in Equation 4.61 relative to the coefficient $1/N$ in Equation 4.55 is to increase the sample variance. This is necessary because the use of $\overline{X}$ constrains the variability of the sum of squares relative to the sum of squares if μ were used. Remember that $\overline{X}$ is derived from X_i, whereas this is not the case when μ is known. When N is large, the effect of not accounting for the loss of one degree of freedom is small, which implies that $\overline{X} \approx \mu$. We've used this argument in Section 4.2.3 to find the bias squared portion of the mean square error of the acvf estimators and will use it again in section 4.7 to find the variance of the acvf estimator portion.

Secondly, when the data are autocorrelated, the variability of the sum of squares is further constrained. Adjacent values in a time series, and hence the mean, vary less the greater the autocorrelation for a given population variance. This reduction is compensated for by $N_e/(N_e - 1)$ in Equation 4.60 with respect to $N/(N-1)$ for the case of uncorrelated data. It is through adjustment of the coefficient of the sum of squares term that improved estimates of the population variance are achieved.

We now have three formulas to apply when want to find the mean and variance of a realization denoted by x_i, $i = 1, \ldots, N$. The sample mean is

$$\bar{x} = \frac{1}{N} \sum_{i=1}^{N} x_i. \tag{4.62}$$

The sample variance is

$$s^2 = \frac{1}{N} \left(\frac{N_e}{N_e - 1} \right) \sum_{i=1}^{N} (x_i - \bar{x})^2 \tag{4.63}$$

which reduces to

$$s^2 = \frac{1}{N-1} \sum_{i=1}^{N} (x_i - \bar{x})^2 \tag{4.64}$$

when the data are uncorrelated.

4.6 Confidence limits for the population mean

Given a data set or realization from which we are trying to gain some insight about the properties of the population time series from which it came, a natural question to ask is, "How representative is the sample mean of the population mean?" Confidence limits for the population mean can be determined from the data set itself. Qualitatively speaking, if the confidence interval is wide, as defined by the confidence limits, the observed mean can be very different than the population mean; if the confidence interval is narrow, the opposite is true. To understand the procedure for obtaining the confidence limits, we consider two computer generated data sets: a realization of white noise and a realization of an AR(1) process. The acf of the process from which the former time series was taken is everywhere zero except at lag 0, while the acf associated with the latter process is everywhere nonzero. We will see the effect of autocorrelation on the width of the confidence interval.

4.6.1 Example of white noise

Let rv X represent normally distributed white noise with standard deviation σ_X and mean μ. Then the distribution of $\bar{X}$, the estimator for the mean, given by

$$\bar{X} = \frac{1}{N} \sum_{i=1}^{N} X_i \tag{4.55}$$

is also normal. We can create a standard normal variable Z according to

$$Z = \frac{\overline{X} - \mu}{\sigma_{\overline{X}}} \tag{4.65}$$

and rewrite Equation 4.65 as

$$\overline{X} = Z\sigma_{\overline{X}} + \mu. \tag{4.66}$$

Therefore, the expression for the $(1-\alpha)\%$ confidence interval applied to the distribution function for $\overline{X}$ is

$$\Pr\left\{ Z\left(\frac{\alpha}{2}\right)\sigma_{\overline{X}} + \mu \leq \overline{X} \leq Z\left(1 - \frac{\alpha}{2}\right)\sigma_{\overline{X}} + \mu \right\} = 1 - \alpha \tag{4.67}$$

where α is the level of significance.

Equation 4.67 can be rearranged to yield the $(1-\alpha)\%$ confidence interval on μ such that

$$\Pr\left\{ \overline{X} - Z\left(1 - \frac{\alpha}{2}\right)\sigma_{\overline{X}} \leq \mu \leq \overline{X} - Z\left(\frac{\alpha}{2}\right)\sigma_{\overline{X}} \right\} = 1 - \alpha. \tag{4.68}$$

The corresponding $(1-\alpha)\%$ confidence limits are

$$\overline{X} \pm Z\left(1 - \frac{\alpha}{2}\right)\sigma_{\overline{X}} \tag{4.69}$$

where, in practice, $\overline{X}$ is replaced by sample mean $\overline{x}$, and $\sigma_{\overline{X}}$ by sample standard deviation $s_{\overline{x}}$.

Figure 4.8 shows a normal white noise realization generated from Equation 4.33 with $X_t = Z_t + \mu$. That is

$$x_t = z_t + \mu, \qquad t = 1, 2, \ldots, N \tag{4.70}$$

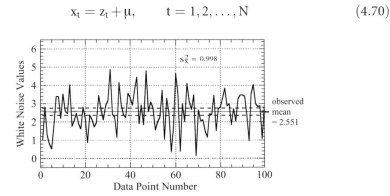

Figure 4.8 A realization of a sequence of 100 values of Gaussian white noise showing the sample mean and variance and the 95% confidence limits for the population mean (dashed lines).

in which a random number generator was used to create samples of white noise from a normal distribution with zero population mean and unit variance so that $\sigma_X = \sigma_Z = 1$. Appendix 4.A discusses a method to create a realization of normal white noise. The population mean of X_t was set to $\mu = 2.5$. Using the formula

$$\bar{x} \pm Z\left(1 - \frac{\alpha}{2}\right)s_{\bar{x}} \tag{4.71}$$

or

$$\bar{x} \pm Z\left(1 - \frac{\alpha}{2}\right)s_x/\sqrt{N - 1} \tag{4.72}$$

with $N = 100$, $s_x = 0.999$, and $\alpha = 0.05$, the 95% confidence limits on the population mean μ are

$$2.551 \pm 1.960 \times 0.999/9.950 = 2.551 \pm 0.197$$

or 2.354 and 2.748, as shown also in Figure 4.8. For this particular realization, both the sample mean and sample variance are very close to their population values. Strictly speaking, a Student's t-distribution should have been used instead of a normal distribution because the population variance was estimated. However, since the number of dof is large ($N - 1 = 99$), the resulting error in the confidence limits for μ is negligible. There was a loss of one dof due to using the sample mean in calculating s_x (you can also review the explanation following Equation 4.61). As is apparent in this example, the loss of one dof could have been ignored with no significant consequence.

4.6.2 Example of a first-order autoregression

Equation 4.35 written in the form $X_t = \alpha(X_{t-1} - \mu) + Z_t + \mu$ is the source of data in this example. The form for creating a realization is

$$x_t = \alpha(x_{t-1} - \mu) + z_t + \mu. \tag{4.73}$$

To initiate the autoregression, let $x_1 = z_1 + \mu$, then let $x_2 = \alpha(x_1 - \mu) + z_2 + \mu$, $x_3 = \alpha(x_2 - \mu) + z_3 + \mu$, and so on. The realization is driven by generating samples of white noise from a normal distribution with zero population mean and unit variance as in the example above. The difference here is that, in addition, an autoregression has to be generated. Figure 4.9 shows a realization of an AR(1) with $\alpha = \rho(1) = 0.90$, $\mu = 2.5$, and $\sigma_Z = 1$. In fact, the time series shown in the figure begins at $t = 25$ in the data generation procedure, the reason being the need to minimize the "beginning effect." The meaning of this term is that the value x_1 is in error because there is no x_0. The error propagates forward in time but decreases with

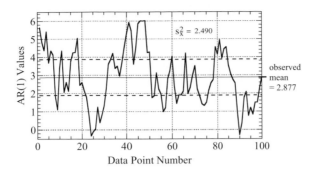

Figure 4.9   A realization of a sequence of 100 values of a first-order autoregressive process with $\rho(1) = 0.9$ showing the sample mean and variance and the 95% confidence limits for the population mean (dashed lines).

each time step. Momentarily, we will see that starting the realization to be used at $t = 25$ is conservative.

We can easily see the difference between a time series that has zero autocorrelation and one that has high autocorrelation. The white noise has closely spaced (in time) ups and downs, while the autocorrelated data show large persistent swings in high values and low values, and only occasional rapid changes in value. The sample mean $\bar{x} = 2.88$ and sample variance $s_x^2 = 2.49$. The latter figure is considerably less than the population value $\sigma_X^2 = 5.26$ calculated from Equation 4.40. We will find out later (Section 4.7.3) that, statistically, the observed variance is reasonable given the population variance and lag 1 correlation.

The procedure to establish the confidence limits for the population mean for white noise in the previous example was straightforward, the reason being that the realization comprised independent data. The equivalent dof, N_e, was equal to the number of data, N. In this example N_e is less than N. We can no longer use N as in Equation 4.72. To find N_e we begin by expanding the summation term in the brackets in Equation 4.54 for an AR(1) process to get

$$\frac{N}{N_e} = 1 + \frac{2}{N}\sum_{k=1}^{N-1}(N-k)\rho_X^k(1)$$

$$= 1 + 2\sum_{k=1}^{N-1}\rho_X^k(1) - \frac{2}{N}\sum_{k=1}^{N-1}k\,\rho_X^k(1). \tag{4.74}$$

Through further expansion, the first summation is a finite geometric series that results in

$$2\sum_{k=1}^{N-1}\rho_X^k(1) = 2\left[\frac{\rho_X(1) - \rho_X^N(1)}{1 - \rho_X(1)}\right]. \tag{4.75}$$

Expansion of the second summation results in

$$-\frac{2}{N}\sum_{k=1}^{N-1}k\,\rho_X^k(1) = -\frac{2}{(1-\rho_X(1))^2}\left[\rho_X^{N+1}(1) - \rho_X^N(1) + \frac{\rho_X(1) - \rho_X^{N+2}(1)}{N}\right].$$

$$(4.76)$$

If N is sufficiently large and $\rho_X(1)$ is somewhat less than one, terms with exponents of N, N + 1, and N + 2 will be small. Within these constraints, Equations 4.75 and 4.76 become, respectively,

$$2\sum_{k=1}^{N-1}\rho_X^k(1) \approx \frac{2\rho_X(1)}{1 - \rho_X(1)}$$

$$(4.77)$$

and

$$-\frac{2}{N}\sum_{k=1}^{N-1}k\,\rho_X^k(1) \approx -\frac{2}{N}\frac{\rho_X(1)}{(1-\rho_X(1))^2}.$$

$$(4.78)$$

With N large, the summation in Equation 4.78 is small compared to the summation in Equation 4.77 so that Equation 4.74 becomes

$$\frac{N}{N_e} \approx 1 + \frac{2\rho_X(1)}{1 - \rho_X(1)}$$

$$(4.79)$$

or, more usefully, the estimated equivalent dof for an AR(1) is

$$N_e \approx N\left(\frac{1 - \rho_X(1)}{1 + \rho_X(1)}\right)$$

$$(4.80)$$

and, from Equation 4.53, the variance of the mean is

$$\sigma_{\bar{X}}^2 \approx \frac{\sigma_X^2}{N}\left[\frac{1 + \rho_X(1)}{1 - \rho_X(1)}\right].$$

$$(4.81)$$

Using Equation 4.23, we find the lag 1 autocorrelation for the realization to be $r_x(1) = 0.77$ in comparison to the population value $\rho_X(1) = \alpha = 0.90$. Replacing $\rho(1)$ by $r_x(1)$ in Equation 4.80 to obtain an estimate of the equivalent dof in the sample yields

$$N_e \approx 100\left(\frac{1 - 0.77}{1 + 0.77}\right)$$
$$\approx 13.$$

$$(4.82)$$

Thus, among the 100 data points there are effectively only 13 independent "pieces of information." Another way to express this result is that only about every eighth point is essentially uncorrelated. Using, again, r_x in place of ρ_X in Equation 4.42 we have $r_x(8) = r_x^8(1) = (0.77)^8 = 0.12$. Apparently, when the autocorrelation drops to about 0.12 in an AR(1) process, another effective dof is created. That we waited until time $t = 25$ to begin the actual realization shown in Figure 4.9 is reasonable, as the "beginning effect" should be negligible. Note, however, that the number of data points required before the beginning effect becomes negligible is directly related to the degree of autocorrelation present.

Because the effective dof, N_e, is so small, a Student's t-distribution must be used in calculating confidence limits. In parallel with Equation 4.72, the formula for the confidence limits on the population mean μ is

$$\overline{x} \pm t\left(1 - \frac{\alpha}{2}\right)s_x / \sqrt{(N_e - 1)} \qquad (4.83)$$

in which the loss of one dof is due to using the sample mean in calculating s_x, just as in the example of white noise. Therefore, the 95% confidence limits are

$$2.88 \pm 2.18 \times 1.58 / \sqrt{12} = 2.88 \pm 0.99$$

or 1.89 and 3.87. The much wider confidence interval in this example than in the white noise example reflects the many fewer degrees of freedom used in estimating the population mean, μ, and population variance, σ_X^2, that, in turn, are a consequence of autocorrelation in the data. As a general statement, we can say that for time series with the same population or sample variance, the greater the serial correlation, the wider the $(1 - \alpha)\%$ confidence interval for the population mean about the sample mean.

4.7 Variance of the acvf and acf estimators

4.7.1 Derivation

In Section 4.2.3 we studied the bias squared portion of the mean square error of the acvf estimators $C(k)$ and $C'(k)$. We concluded that, in anticipation of the results of this section, the preferred estimator for the acvf is the biased estimator $C(k)$ because of the smaller mean square error with increasing lag. In this section we derive expressions for the variance of the biased and unbiased acvf estimators, use them to find the variance of the acvf and acf estimators for white noise and a first-order autoregression, and then consider an example of each. To make the derivations tractable we consider the case in which the population mean $\mu_X = 0$. In addition, we assume the number of data, N, in a realization is sufficiently large that the distribution of rv $\overline{X}$ is narrow enough that any value in the distribution is approximately zero. With these conditions, only the first term on the right-hand sides of Equations and 4.18 contributes significantly to the expectation on the left-hand side.

For C(k) we have

$$
\begin{aligned}
\mathrm{Var}[C(k)] &= E\big[(C(k) - E[C(k)])^2\big] \\
&= E\left[\left(C(k) - \left(\frac{N - |k|}{N}\right)\gamma(k)\right)^2\right] \\
&= E[C^2(k)] - \left(\frac{N - |k|}{N}\right)^2 \gamma^2(k)
\end{aligned}
\tag{4.84}
$$

where C(k) is the biased acvf estimator and N is the number of data in a realization. Expanding the first term on the right yields

$$
\begin{aligned}
E[C^2(k)] &= E\left[\frac{1}{N}\sum_{i=0}^{N-|k|-1} X_i X_{i+|k|} \frac{1}{N}\sum_{i=0}^{N-|k|-1} X_i X_{i+|k|}\right] \\
&= \frac{1}{N^2}\sum_{i=0}^{N-|k|-1}\sum_{j=0}^{N-|k|-1} E\big[X_i X_{i+|k|} X_j X_{j+|k|}\big].
\end{aligned}
\tag{4.85}
$$

As shown in Bendat and Piersol (1966, p. 94), the expectation of the product of four random variables that follow a four-dimensional normal distribution with possibly different nonzero means is given by

$$
\begin{aligned}
E[X_1 X_2 X_3 X_4] &= E[X_1 X_2]\ E[X_3 X_4] + E[X_1 X_3]\ E[X_2 X_4] \\
&\quad + E[X_1 X_4]\ E[X_2 X_3] - 2\mu_{X_1}\mu_{X_2}\mu_{X_3}\mu_{X_4}.
\end{aligned}
\tag{4.86}
$$

Therefore, for our case with all means being zero

$$
\begin{aligned}
E[C^2(k)] &= \frac{1}{N^2}\sum_{i=0}^{N-|k|-1}\sum_{j=0}^{N-|k|-1}\Big\{E[X_i X_{i+|k|}]\ E[X_j X_{j+|k|}] \\
&\quad + E[X_i X_j]\ E[X_{i+|k|} X_{j+|k|}] + E[X_i X_{j+|k|}]\ E[X_{i+|k|} X_j]\Big\}
\end{aligned}
\tag{4.87}
$$

and

$$
\begin{aligned}
\mathrm{Var}[C(k)] &= \frac{1}{N^2}\sum_{i=0}^{N-|k|-1}\sum_{j=0}^{N-|k|-1}\Big\{\gamma^2(k) + \gamma^2(j - i) + \gamma(j + k - i)\,\gamma(j - k - i)\Big\} \\
&\quad - \left(\frac{N - |k|}{N}\right)^2 \gamma^2(k).
\end{aligned}
\tag{4.88}
$$

Similarly, the variance of the unbiased estimator is

$$Var[C'(k)] = \frac{1}{(N-|k|)^2} \sum_{i=0}^{N-|k|-1} \sum_{j=0}^{N-|k|-1} \{\gamma^2(k) + \gamma^2(j-i) + \gamma(j+k-i)\,\gamma(j-k-i)\}$$
$$- \gamma^2(k). \tag{4.89}$$

Equations 4.88 and 4.89 are the equations for the variance of the biased and unbiased acvf estimators, respectively, for any stationary normal random process, and require knowledge of the process acvf. In comparing the two variances, we can see that as |k| increases, their ratio $Var[C'(k)]/Var[C(k)]$ will also increase, indicating the increasing variability of estimator $C'(k)$ relative to estimator $C(k)$. To repeat the conclusion in Section 4.2.3, "It is thought that for most acvfs the effect of increasing variance of $C'(k)$ overwhelms the bias squared effect associated with $C(k)$ (Jenkins and Watts, 1968, pp. 179–180)." The result was our recommendation to use the biased estimator $C(k)$.

Now let us consider the case when $k=0$ wherein Equations 4.88 and 4.89 reduce to

$$Var[C(0)] = Var[C'(0)] = \frac{2}{N^2} \sum_{i=0}^{N-1} \sum_{j=0}^{N-1} \gamma^2(j-i). \tag{4.90}$$

The procedure following Equation 4.50 can be applied to Equation 4.90 to yield the variance of the variance of a stationary random process, namely

$$Var[S_X^2] = \frac{2\sigma_X^4}{N} \left[1 + \frac{2}{N} \sum_{k=1}^{N-1} (N-k)\,\rho_X^2(k) \right] \tag{4.91}$$

which parallels the variance of the mean given by Equation 4.52.

4.7.2 White noise

For white noise and for $|k| > 0$, both $\gamma^2(k)$ and the product $\gamma(j+k-i)\,\gamma(j-k-i)$ in Equations 4.88 and 4.89 are always zero. Thus, for the biased acvf estimator

$$Var[C(k)] = \frac{1}{N^2} \sum_{i=0}^{N-|k|-1} \sum_{j=0}^{N-|k|-1} \gamma^2(j-i)$$

in which the argument of the double summation is nonzero only when $i=j$. This reduces to

$$Var[C(k)] = \frac{N-|k|}{N^2} \sigma_X^4 \tag{4.92}$$

while for the unbiased estimator

$$\text{Var}[C'(k)] = \frac{1}{N - |k|}\sigma_X^4. \tag{4.93}$$

Note that the variance of C(k) decreases with increasing lag while the variance of C'(k) increases with increasing lag. However, if $|k| \ll N$ then

$$\text{Var}[C(k)] \approx \text{Var}[C'(k)] \approx \frac{\sigma_X^4}{N}, \quad 0 < |k| \ll N. \tag{4.94}$$

By dividing C(k) and C'(k) by $\gamma(0) = \sigma_X^2$, the variance of the acf estimator becomes

$$\text{Var}[R(k)] \approx \text{Var}[R'(k)] \approx \frac{1}{N}, \quad 0 < |k| \ll N. \tag{4.95}$$

It should be remarked that if the sample mean had been used in the derivation, the variance of the acf and acvf estimators would be somewhat larger than those given due to the added variability of the sample mean. The procedures to find the variance of the acvf estimator when the sample mean is used and the variance of the acf estimator when it is standardized with C(0) are very complex. The problem has been attacked by Anderson (1942) using a *circular* acf, which is a consequence of considering periodic processes. His results were cited in Section 4.4.4.

For white noise, $\rho_X^2(k > 0) = 0$ in Equation 4.91, so that

$$\text{Var}[S_X^2] = \frac{2\sigma_X^4}{N}. \tag{4.96}$$

If we generate white noise with population variance σ_X^2, the sample variance should reasonably lie within two standard deviations about the population variance, so that from Equation 4.96

$$(1 - 2\sqrt{2/N})\sigma_X^2 < s_x^2 < (1 + 2\sqrt{2/N})\sigma_X^2.$$

Let's see if this is true for the white noise example in Section 4.6.1 and shown in Figure 4.8. In that example, $\sigma_X^2 = 1$ and $N = 100$, so the limits of two standard deviations about the population variance are 1 ± 0.28 or 0.72 and 1.28. Since the observe variance $s_x^2 = 0.998$, it is clearly within these limits. In fact, it is well within one standard deviation of the population variance.

4.7.3 First-order autoregression

A procedure similar to that beginning with Equation 4.74 can be followed to show that for a first-order autoregressive process, Equation 4.91 reduces to

$$\text{Var}[S_X^2] \approx \frac{2\sigma_X^4}{N}\left[\frac{1 + \rho_X^2(1)}{1 - \rho_X^2(1)}\right]. \tag{4.97}$$

Equation 4.97 shows that, as expected, the more autocorrelated the data, the more variable are estimates of the process variance for a given N. This is because as the serial correlation increases the degrees of freedom (dof) decrease.

We apply Equation 4.97 to the example of a first-order autoregression in Section 4.6.2 and shown in Figure 4.9. As in the white noise example above, we should reasonably expect the observed variance $s_x^2 = 2.49$ to lie within two standard deviations about the population variance $\sigma_X^2 = 5.26$. From Equation 4.97 these limits are

$$5.26 \pm 2\sqrt{2/N} \times 5.26 \times \left[\frac{1 + \rho_X^2(1)}{1 - \rho_X^2(1)}\right]^{1/2} = 5.26 \pm 0.28 \times 5.26 \times 3.09 = 0.71 \text{ and } 9.80$$

in which $N = 100$ and $\rho_X(1) = 0.9$. The observed variance is well within two standard deviations of σ_X^2. The respective limits for one standard deviation are 2.98 and 7.54, so we see that s_x^2 lies between the lower one and two standard deviation limits. Thus the variance of the realization is consistent with the population variance and lag 1 autocorrelation.

Appendix 4.A   Generating a normal random variable

Let X_1, X_2, X_3, ..., X_n be a sequence of n independently and identically distributed (iid) random variables each having expectation μ and variance σ^2. If we form the sum rv $S = X_1 + X_2 + \cdots + X_n$, the central limit theorem tells us that the distribution of rv

$$Z = \frac{S - n\mu}{\sigma\sqrt{n}} \tag{4.A.1}$$

approaches the standard normal distribution $N(0,1)$ as n tends to ∞. $N(0,1)$ is a Gaussian or normal distribution with zero mean and unit variance. The denominator on the right-hand side of Equation 4.A.1 is a consequence of iid; that is, the variance of S is n times the variance of each rv or Var $[S] = n\sigma^2$.

To specify a population variance v^2 other than unit variance, scale (i.e., multiply) the right-hand side of Equation 4.A.1 by v, the standard deviation. To specify a population mean d other than zero, add the desired mean to the right-hand side of Equation 4.A.1. To change both the population mean and variance of Equation 4.A.1 from zero and one to d and v^2, respectively, rewrite Equation 4.A.1 as

$$Z^* = \frac{S - n\mu}{\sigma\sqrt{n}}v + d. \tag{4.A.2}$$

Equation 4.A.2 approaches the normal distribution $N(d, v^2)$ as n tends to ∞.

Now consider an application of Equation 4.A.1. Let us say we have $n = 50$ random variables, X_1, X_2, ..., X_{50}, each having a uniform distribution between zero and one. Practically any computer that can be used for scientific purposes has a command to generate a random number from a uniform distribution. We withdraw a sample from each of the 50 distributions and add them to create a sample of the sum rv S. Then, following Equation 4.A.1, we calculate a sample of rv

$$Z = \frac{S - 25}{\sqrt{50/12}}. \qquad (4.A.3)$$

Numerical values 25 and 12 on the right-hand side of Equation 4.A.3 were obtained from the statistical properties of a uniform distribution. The mean is $(a + b)/2$, where a is its lower limit and b its upper limit, and the variance is $(b-a)^2/12$. In Equation 4.A.3, $b = 1$ and $a = 0$. The derivation of the variance of a uniform distribution was problem 7 in Chapter 1.

Random variable Z has an approximate normal distribution with expected value zero and expected variance one. From Equation 4.A.3 we see that by computing successive samples of S and then Z we can simulate a normal time series. The actual distribution of Z will have a mean different from zero (or d, if Equation 4.A.2 is used) and a variance different from one (or v^2, if Equation 4.A.2 is used). The larger n is, on average, the closer the mean and variance will be to their expected values.

A respectable normal distribution can be obtained from a uniform distribution with n as small as 12. An improved normal distribution can be obtained using $n = 25$ and greater. Of course, the tails of the distribution will always be truncated.

Problems

1 (a) Show that the variance of the sum of two independent random variables X and Y is equal to the sum of the variances of the individual random variables.

(b) Based on your answer for two random variables X and Y, can a more general statement be made? If so, what is it?

2 (a) Using the biased estimator (i.e., where the coefficient is 1/N), calculate the sample acvf at lag 3 of the mini-time series given by

$$x_1 = 3 \qquad x_2 = 7 \qquad x_3 = 9 \qquad x_4 = 5 \qquad x_5 = 2 \qquad x_6 = -2$$

(b) Calculate the sample autocorrelation function at lag 3.

3 Show that the coefficient α in the AR(1) process

$$X_t - \mu_X = \alpha(X_{t-1} - \mu_X) + Z_t$$

is identical to the value of the population autocorrelation function of the process at lag 1.

4 (a) Given the sequence of six random numbers z_n shown below from a normal distribution $N(0, s^2)$, generate a sample sequence of x_n of an AR(1) process where $r(1) = 0.5$.

n	z_n	x_n
0	1	
1	3	
2	−1	
3	2	
4	−3	
5	−2	

(b) What is the earliest time step, n', at which it would be reasonable to say that the "beginning effect" is negligible? Explain your choice.

5 Derive the formula for the variance of the AR(1) process

$$X_t - \mu_X = \alpha(X_{t-1} - \mu_X) + Z_t$$

6 (a) What is the equation for the population acvf for a stationary time series using the expectation operator? Define each quantity.

(b) Write down the same equation except for a nonstationary time series and explain why they are different.

7 Show that the variance of the AR(2) process

$$X_t - \mu_X = \alpha_1(X_{t-1} - \mu_X) + \alpha_2(X_{t-2} - \mu_X) + Z_t$$

in which

$$\gamma_X(k) = \alpha_1\gamma_X(k-1) + \alpha_2\gamma_X(k-2), \quad k \geq 1$$

can be given by

$$\sigma_X^2 = \frac{\sigma_Z^2}{\left[1 - \alpha_1^2\left(\dfrac{1+\alpha_2}{1-\alpha_2}\right) - \alpha_2^2\right]}.$$

Include any explanations that will clarify your derivation.

8 Write a computer program to compute the sample acvf using the formula

$$c(k) = (1/N) \sum_{t=1}^{N-|k|} (x_t - \bar{x})(x_{t+|k|} - \bar{x}), \quad k = 0, \pm 1, \pm 2, \dots$$

9 In this problem you will explore the acvf and acf for an analytical function and realizations from three different random processes. For each series in parts (a)–(d) below, compute the acvf and acf for a series 100 points long with a maximum lag of 25 points. For parts (c) and (d), generate 50 to 100 points to stabilize the process before obtaining the sample data set.

(a) A sinusoid given by

$$A_1 \cos(2\pi f_1 r \Delta t + \phi_1) + B_1$$

where $\phi_1 \neq 0$, B_1 is a nonzero constant, $f_1 = m/(100\Delta t)$ where $4 < m < 20$, and $r = 0, 1, 2, \dots, 99$.

(b) A white noise process given by

$$X_t = Z_t$$

where Z_t is Gaussian white noise with $E[Z_t] = 0$ and $Var[Z_t] = \sigma_Z^2 = 4$. [Note that Z_t should be an independent sequence for each realization in (b), (c), and (d).]

(c) An AR(1) process given by

$$X_t = \alpha_1 X_{t-1} + Z_t$$

where Z_t is as defined in (b) and, in general, $0 < \alpha_1 < 1$. For this particular case, use $0.5 < \alpha_1 < 1$ in order to easily observe the autocorrelation in the data.

(d) An AR(2) process given by

$$X_t = \alpha_1 X_{t-1} + \alpha_2 X_{t-2} + Z_t$$

where Z_t is as defined in (b), $\alpha_1 = 0.5$, and $\alpha_2 = -0.7$.

(e) Plot the data sets in (a)–(d). On each plot use a solid line to show the time series, a dashed line to show the sample mean, and a dotted line to show the population mean. Also, on each plot show the numerical values of the sample and population means and variances.

(f) Plot the sample acf (solid line) and the population acf (dashed line) out to lag 25 for each of the four time series, with one sample acf and the associated population acf per graph. For time series (a), assume the "population" acf can be represented by an infinite continuous sinusoid of the form described. For time series (d), the population acf is given by

$$\rho(k) = \frac{R^{|k|} \sin(2\pi f_0 k + \phi_0)}{\sin\phi_0}$$

where

$$R = \sqrt{-\alpha_2}$$

$$\cos(2\pi f_0) = |\alpha_1|/(2R)$$

$$\tan\phi_0 = \left[\frac{1 + R^2}{1 - R^2}\right] \tan(2\pi f_0)$$

(g) For the data in (a), explain how the shape of the acf would have changed if $1/(N - |k|)$ had been used instead of $1/N$ in its formula.

(h) Decide whether the sample mean you calculated for (b) is "reasonable" by determining the 95% confidence interval for the sample mean.

(i) Decide whether the sample variance you calculated for (b) is "reasonable" by determining the 95% confidence interval for the sample variance. Use Equation 4.96 and assume that the sample variances have a Gaussian distribution.

10 Give three examples of nonmeteorological time series and identify whether those series would likely be stationary or nonstationary and why.

11 Consider the estimator for the mean given by

$$\overline{X} = (1/N) \sum_{n-1}^{N} X_n$$

(a) Given that $\mu = E[X]$, show that the above is an unbiased estimator of the population mean μ for the X-stationary random process in which the random variables X_n are independent.

(b) How, if it all, would the above estimator for the mean change if the random variables X_n were dependent?

(c) How, if it all, would the above estimator for the mean change if the random variables X_n were nonstationary?

12 Let us say that we have fitted a zero mean AR(1) process

$$X_t = \alpha X_{t-1} + Z_t$$

to a realization of data comprising 500 hPa heights. Notation is standard and α is positive. We then wish to use this stochastic model to forecast values of X at times $t + 1$, $t + 2$, and so on, where t is the current time. Thus,

$$\hat{X}_t(1) = \alpha X_t$$
$$\hat{X}_t(2) = \alpha \hat{X}_t(1) = \alpha^2 X_t$$
$$\hat{X}_t(3) = \alpha \hat{X}_t(2) = \alpha^3 X_t$$

$$\vdots$$

where the karat ($\wedge$) indicates the forecast. The successive forecast errors are given by

$$e_t(1) = \hat{X}_t(1) - X_{t+1} = \alpha X_t - X_{t+1}$$
$$e_t(2) = \hat{X}_t(2) - X_{t+2} = \alpha^2 X_t - X_{t+2}$$
$$e_t(3) = \hat{X}_t(3) - X_{t+3} = \alpha^3 X_t - X_{t+3}$$

$$\vdots$$

(a) Show that the error variance for the k-th forecast step ahead is

$$\mathrm{Var}[e_t(k)] = \left(1 - \alpha^{2k}\right)\sigma_X^2$$

(b) Does the error variance increase or decrease with increasing k?

(c) What is the limiting value of the error variance as k increases indefinitely? Why is this so?

(d) Show that

$$\mathrm{Cov}[e_t(k), e_t(k+1)] = \alpha\left(1 - \alpha^{2k}\right)\sigma_X^2$$

(e) Are the errors associated with the forecasts serially correlated?

(f) What is the limiting value of the error covariance as k increases indefinitely? Why is this so?

(g) Is the time series of errors stationary or nonstationary? Explain.

References

Anderson, R.L. (1942) Distribution of the serial correlation coefficient. *Ann. Math. Stat.*, **13**, 1–13.

Bendat, J.S., and Piersol, A.G. (1966) *Measurement and Analysis of Random Data.* John Wiley & Sons, Inc., New York.

Box, G.E.P., and Jenkins, G.M. (1970) *Time Series Analysis Forecasting and Control.* Holden-Day, San Francisco, CA.

Jenkins, G.M., and Watts, D.G. (1968) *Spectral Analysis and Its Applications.* Holden-Day, San Francisco, CA.

Ljung, G.M., and Box G.E.P (1978) On a measure of lack of fit in time series analysis models. *Biometrika*, **65**, 297–303.

5
Lagged-product spectrum analysis

5.1 The variance density spectrum

A periodogram yields a high-resolution spectrum of the variance in a time series. For some physical situations a periodogram is exactly what is needed to identify periodicities that are thought to exist. An example is the daily cycle of temperature that was studied in Chapters 1 and 3. In other situations, a physical phenomenon may have considerable variation from one occurrence to the next. An example is the time series of surface wind speed associated with Chinooks (USA) or Foehns (Europe). The periodograms of wind speed from successive events may be very noisy-looking, such that it is difficult to draw any conclusion of the general structure of the variance of wind speed versus frequency. For this kind of situation there are three options to consider. One is to smooth each periodogram by applying a running average of harmonic variances as was done in Section 1.4.5. Another is to average an ensemble of event spectra harmonic-by-harmonic, similar to the procedure in Section 1.5.3. The third is to obtain inherently smooth spectra of the events through *lagged-product spectrum analysis*. In this approach, the spectrum is the Fourier transform of the autocovariance function (comprising lagged products) that was derived in Chapter 4. The spectrum that results from the lagged-product method is called a *variance density spectrum* because it has units associated with variance per unit bandwidth. Common dimensions of bandwidth are cycles or radians per unit of time. As we know, the periodogram has units of variance only.

The issue, though, is not variance or variance density versus frequency but the degree of smoothing in either type of spectrum. The lagged-product method can offer "instant" smoothing via Fourier transformation of the autocovariance function

Time Series Analysis in Meteorology and Climatology: An Introduction, First Edition. Claude Duchon and Robert Hale.
© 2012 John Wiley & Sons, Ltd. Published 2012 by John Wiley & Sons, Ltd.

(acvf). This happens in two ways. In the first way, simply truncating the acvf at some lag less than $(N - 1)$ and Fourier transforming the truncated acvf to the frequency domain produces a spectrum that is smoother than a periodogram of the same data set. Truncating an acvf is equivalent to multiplying the complete acvf by a rectangular window. The narrower the window the smoother the spectrum. As we learned in Section 2.4 (see Figure 2.7), the Fourier transform of a rectangular function is a diffraction function with positive and negative side lobes. Recalling that Fourier transformation of multiplication in the time domain leads to convolution in the frequency domain, the smoothing that occurs in the variance density spectrum will include the undesirable effects of the negative and positive side lobes. To suppress these effects, instead of multiplying the complete acvf by a rectangular window, some other window with a more gradual approach from zero lag to the positive and negative lag where the window has zero value can be applied. The result is an "improved" smooth variance density spectrum. The last section of this chapter deals with a cosine window (as opposed to a rectangular window) and the consequent effects on smoothing of the variance density spectrum.

Another reason to choose the lagged-product method instead of the periodogram method is because in certain problems the autocovariance (or autocorrelation) function plays a significant role and has to be computed anyway. For example, in turbulence theory, the "Lagrangian integral time scale" (Tennekes and Lumley, 1972, p. 46), a measure of the time a variable such as wind speed is correlated with itself, is obtained by integration of the autocorrelation function. If this is an important quantity to know, and if there is further interest in the spectrum of turbulence, which is usually the case, it is then a simple matter to Fourier transform the acvf to obtain the spectrum of turbulence. Alternatively, one might be given a spectrum and wish to know the time (or space) correlation structure of the signal that produced it. This requires back-transforming the variance density spectrum to get the acvf.

We now provide more detail to the procedure involved in lagged-product spectrum analysis. The first step is to multiply the *raw* or *unwindowed acvf* by a *lag window* to yield a *windowed acvf*. Then the windowed acvf is Fourier transformed to the frequency domain to yield the *smoothed spectrum*. This is the procedure shown in Figure 5.1a. Equivalently, the smoothed spectrum can be viewed as the convolution of the Fourier transform of the unwindowed or raw acvf with the Fourier transform of the lag window. The result of the former transformation is the *unsmoothed* or *raw spectrum*; the result of the latter transformation is the *spectrum window*. This procedure is shown in Figure 5.1b. Thus it is necessary to distinguish between a raw (or unwindowed) acvf and a windowed acvf and a raw (or unsmoothed) spectrum and a smoothed spectrum. A number of lag windows have been designed to effect a varying degree of spectrum smoothing. The smoothing is usually done to the acvf (as in Figure 5.1a) rather than to the raw spectrum (as in Figure 5.1b) simply as a matter of convenience.

From a theoretical viewpoint, a variance density spectrum is required when the underlying process for some geophysical phenomenon is continuous in time and

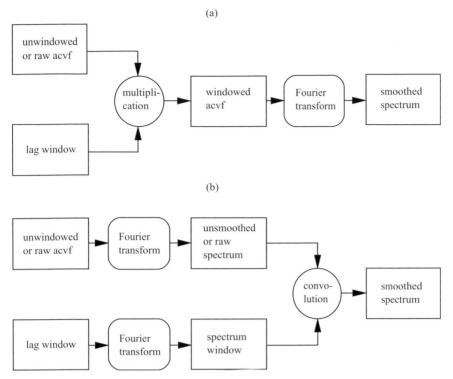

(a)

(b)

Figure 5.1 (a) The smoothed spectrum via the Fourier transform of the windowed acvf. (b) The same smoothed spectrum via convolution of the spectrum window with the unsmoothed or raw spectrum.

aperiodic, which is typically the case. Then the spectrum perforce must be continuous in frequency, with its relationship to the acvf given by

$$\Gamma(f) = \int_{-\infty}^{+\infty} \gamma(u) \exp(-i2\pi fu) \, du \qquad (5.1)$$

where $\Gamma(f)$ and $\gamma(u)$ are the population variance density spectrum and population acvf, respectively. The inverse Fourier transform is

$$\gamma(u) = \int_{-\infty}^{+\infty} \Gamma(f) \exp(i2\pi fu) \, df. \qquad (5.2)$$

The Fourier transform pair comprising Equation 5.1 and Equation 5.2 is known as the Wiener–Khintchine relation (Koopmans, 1974, pp. 33–34). In consideration of the dimensions associated with u and $\gamma(u)$ in Equation 5.1 it is evident that $\Gamma(f)$ must have dimensions associated with variance per unit bandwidth, that is, variance density. To compute variance it is necessary to integrate over some bandwidth. This is analogous to the familiar probability density function from which probability is obtained by integrating over some specified range of the independent variable.

Equations 5.1 and 5.2 provide the motivation for the next section. Instead of dealing with integrals and analog data, we will deal with summations and digital data to find the relation between the variance density spectrum and the autocovariance function. As a preliminary step, we recall the Fourier variance spectrum (Equation 1.69) and divide it by the separation between harmonic frequencies, that is, bandwidth $1/(N\Delta t)$, to obtain

$$C(f) = \frac{\overset{*}{S'}(f) \times S'(f)}{1/(N\Delta t)}, \qquad -1/(2\Delta t) \le f \le 1/(2\Delta t) \tag{5.3}$$

where N is the number of data, Δt the sampling interval, $S'(f)$ the complex Fourier amplitude coefficient at frequency f, and $\overset{*}{S'}(f)$ its conjugate. The purpose of the primes on the right-hand side of Equation 1.69 was to indicate a two-sided spectrum because of the need to distinguish between two-sided and one-sided spectra in Chapter 1. Since all the mathematical development in this chapter is involved with two-sided spectra, we can dispense with use of a prime attached to $C(f)$ in Equation 5.3 and at the same time create unique notation for a variance density spectrum. While $C(f)$ is an ordinary mathematical variable in Equation 5.3, in Section 5.3 we will consider $C(f)$ to be also a random variable. This will enable us to understand certain properties of variance density spectra for random processes, the results of which we will use in Section 5.4.

Equation 5.3 shows that a variance density spectrum and a variance spectrum differ by the term for the bandwidth, $1/(N\Delta t)$. To obtain the total variance in spectrum $C(f)$, products of $C(f_i)$ and the frequency separation $f_{i+1} - f_i = \Delta f_i$ between adjacent estimates need to be summed over the range in frequency $-1/(2\Delta t)$ to $1/(2\Delta t)$, that is, the principal part of the aliased spectrum. In short, numerical integration must be performed.

5.2 Relationship between the variance density spectrum and the acvf

In this section we derive the relationship between the variance density spectrum and the autocovariance function. The derivation begins by substituting Equation 1.65 and its complex conjugate into Equation 5.3 to obtain

$$C(f) = (\Delta t/N) \sum_{n=0}^{N-1} x_n\, e^{i2\pi f n \Delta t} \sum_{n'=0}^{N-1} x_{n'}\, e^{-i2\pi f n' \Delta t},$$
$$-1/(2\Delta t) \le f \le 1/(2\Delta t) \tag{5.4}$$

where x_n and $x_{n'}$ are, for ease in the derivation, departures from the sample mean. Rearranging Equation 5.4 leads to

$$C(f) = (\Delta t/N) \sum_{n'=0}^{N-1} \sum_{n=0}^{N-1} x_{n'} \, x_n \, e^{-i2\pi f(n'-n)\Delta t}. \tag{5.5}$$

Figure 5.2 shows the area of summation. Diagonals from the lower left to the upper right represent lines for which $n' - n = k$ where k is a constant. Such lines contain the products in an acvf calculation for lag k. Thus, by employing the coordinate transformation $k = n' - n$ and $m = n$, k becomes lag number and m is the number of products for a given lag. The result of the transformation is that products along a diagonal line in Figure 5.2 become products along a horizontal line in Figure 5.3. For example, products along the main diagonal (corner C to corner B) in Figure 5.2 become the same products along the horizontal line $k = 0$ (corner C to corner B) in Figure 5.3. Products in diagonal lines above the main diagonal in Figure 5.2 become products in horizontal lines above $k = 0$ in Figure 5.3, and, similarly, products in diagonal lines below the main diagonal in Figure 5.2 become products in horizontal lines below $k = 0$ in Figure 5.3. Thus, Equation 5.5 can be written

$$C(f) = (\Delta t/N) \sum_{k} \sum_{m} x_{m+k} \, x_m \, e^{-i2\pi fk\Delta t}. \tag{5.6}$$

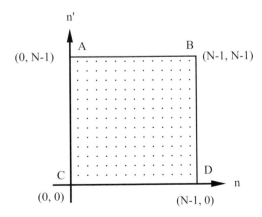

Figure 5.2 The area of summation in Equation 5.5.

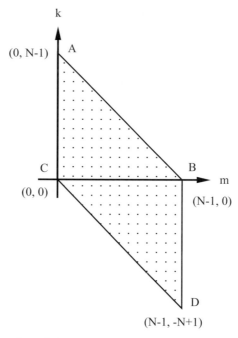

Figure 5.3   Transformation of Figure 5.2 to provide the two summations in Equation 5.7.

The specific limits are determined from Figure 5.3, which shows the field of products, now in the k, m coordinate system. We have

$$C(f) = \Delta t \sum_{k=0}^{N-1} \left(\frac{1}{N} \sum_{m=0}^{N-k-1} x_{m+k}\, x_m \right) e^{-i2\pi fk\Delta t}$$

$$+ \Delta t \sum_{k=-(N-1)}^{-1} \left(\frac{1}{N} \sum_{m=-k}^{N-1} x_{m+k}\, x_m \right) e^{-i2\pi fk\Delta t}. \tag{5.7}$$

The first term on the right-hand side is the sum over the upper part of Figure 5.3 and the second term is the sum over the lower part. The quantities inside the parentheses are biased autocovariance functions. That is, in accord with Equation 4.11 and recalling that we are dealing with departures from the sample mean,

$$\frac{1}{N} \sum_{m=0}^{N-k-1} x_{m+k}\, x_m = c(k) = c(-k), \quad k = 0,\ 1,\ \ldots,\ N-1$$

and

$$\frac{1}{N} \sum_{m=-k}^{N-1} x_{m+k}\, x_m = c(k) = c(-k), \quad k = -(N-1),\ -(N-2),\ \ldots,\ -1$$

so that

$$C(f) = \Delta t \sum_{k=-(N-1)}^{N-1} c(k)\, e^{-i2\pi fk\Delta t}, \qquad -1/(2\Delta t) \le f \le 1/(2\Delta t). \qquad (5.8)$$

Equation 5.8 shows the relation between the variance density spectrum and the acvf. We can expand Equation 5.8 to take advantage of the fact that the acvf is an even function, thereby leading to

$$C(f) = \Delta t \sum_{k=-(N-1)}^{N-1} c(k)\, \cos(2\pi fk\Delta t)$$

$$= \Delta t \left[c(0) + 2 \sum_{k=1}^{N-1} c(k)\, \cos(2\pi fk\Delta t) \right]. \qquad (5.9)$$

Equation 5.9 is an appropriate formula to calculate an analog variance density spectrum from an autocovariance function. We can go one step further and calculate a periodogram from Equation 5.9 by letting $f = m/(N\Delta t)$ and multiplying both sides by $1/(N\Delta t)$. The result is

$$C'_m = \frac{1}{N} \left[c(0) + 2 \sum_{k=1}^{N-1} c(k)\, \cos(2\pi km/N) \right]. \qquad (5.10)$$

Equation 5.10 will yield the same periodogram variances as in Equation 1.63 for a two-sided periodogram using Fourier coefficients. Recall that the amplitude A_0 of the harmonic at $m=0$ is the mean of the time series. Consequently, there is no variance at this harmonic; that is, $C'_0 = 0$. By setting $m=0$ in Equation 5.10 we see that the sum of $c(k)$ over its entire range, that is, from $-(N-1)$ to $(N-1)$, must be zero. The proof involves rearranging the terms in the acvf to form sums of deviations of the data from the sample mean, each sum being zero. This is another interesting exercise for the reader to consider. In summary, we see that just as the Fourier transform of the acvf results in a variance density spectrum, with a simple modification, we also can obtain a periodogram.

Equation 5.8 is one member of the Fourier transform pair between an analog variance density spectrum and an acvf. We anticipate the other half to be

$$c(k) = \int_{-1/(2\Delta t)}^{1/(2\Delta t)} C(f)\, e^{i2\pi fk\Delta t}\, df, \quad |k| \le N - 1. \qquad (5.11)$$

Verification is given in Appendix 5.A. In practice, we are unable to perform the integration required in Equation 5.11, so that if we wish to obtain the acvf from the spectrum (i.e., compute the inverse or back transform of the spectrum) we need to use summation instead of integration. Appendix 5.B shows that we can obtain the acvf by applying a Fourier transform to the spectrum derived from the fully-lagged acvf. Thus,

$$c(k) = \frac{1}{2N\Delta t} \sum_{n=-(N-1)}^{N} C(f_n)\, e^{i2\pi f_n k\Delta t} \tag{5.12}$$

where $f_n = n/(2N\Delta t)$. We observe that there are 2N frequencies and the separation between frequencies is $1/(2N\Delta t)$, one-half that we might have expected.

In summary, there is a strong parallel between periodogram analysis and Equations 5.8 and 5.12. The acvf $c(k)$ can be thought of as a "time series," so that its Fourier transformation yields its spectral decomposition. The acvf contains squared and cross-product terms and its Fourier transform leads to a variance density spectrum. In periodogram analysis, the Fourier transform of a time series of data leads to a Fourier amplitude spectrum. The product of the amplitude spectrum and its complex conjugate yields a variance spectrum, as was demonstrated in Chapter 1.

5.3 Spectra of random processes

5.3.1 Population spectrum

As stated in Section 5.1, we now treat $C(f)$ as a random variable. Furthermore, we consider the upper-case version of $c(k)$, namely $C(k)$, to also be a random variable. This allows us to take the expectation of Equation 5.8, yielding

$$E[C(f)] = \Delta t \sum_{k=-(N-1)}^{N-1} E[C(k)]\, e^{-i2\pi fk\Delta t}$$

$$= \Delta t \sum_{k=-(N-1)}^{N-1} E\left[\frac{1}{N}\sum_{t=0}^{N-|k|-1}\left(X_{t+|k|}-\overline{X}\right)\left(X_t-\overline{X}\right)\right] e^{-i2\pi fk\Delta t},$$

$$-1/(2\Delta t) \le f \le 1/(2\Delta t). \tag{5.13}$$

Using Equation 4.17, we have the result that

$$E[C(f)] = \Delta t \sum_{k=-(N-1)}^{N-1}\left(1-\frac{|k|}{N}\right)\left(\gamma(k)+\text{Var}\left[\overline{X}\right]\right) e^{-i2\pi fk\Delta t}$$

$$-\Delta t \sum_{k=-(N-1)}^{N-1}\frac{1}{N}\sum_{t=0}^{N-|k|-1}\left\{E\left[\left(\overline{X}-\mu\right)\left(X_{t+|k|}-\mu\right)\right]\right.$$

$$\left.+E\left[\left(\overline{X}-\mu\right)\left(X_t-\mu\right)\right]\right\} e^{-i2\pi fk\Delta t}. \tag{5.14}$$

Thus the expected spectrum is itself a function of the number of data N. This is true even if the unbiased acvf estimator had been used. The definition of the population spectrum requires taking the limit of an increasing number of data. Accordingly,

$$\Gamma(f) \equiv \lim_{N \to \infty} E[C(f)] = \Delta t \sum_{k=-\infty}^{\infty} \gamma(k) \, e^{-i2\pi fk\Delta t}, \quad -1/(2\Delta t) \le f \le 1/(2\Delta t). \tag{5.15}$$

The inverse Fourier transform is

$$\gamma(k) = \int_{-1/(2\Delta t)}^{1/(2\Delta t)} \Gamma(f) \, e^{i2\pi fk\Delta t} \, df, \quad k = 0, \pm 1, \pm 2 \ldots. \tag{5.16}$$

Equations 5.15 and 5.16 are the digital equivalent of the Wiener–Khintchine relations (Equations 5.1 and 5.2). Both the analog and digital forms require record lengths that tend to infinity to obtain the population spectrum.

5.3.2 Spectra of linear processes

In Chapter 4 we showed that

$$\gamma_X(k) = \sum_{m=0}^{\infty} \sum_{n=0}^{\infty} h_m \, h_n \, \gamma_Z(k - n + m) \tag{4.28}$$

relates the acvf $\gamma_X(k)$ of output process X_t to the acvf $\gamma_Z(k)$ of input process Z_t after the latter passes through a linear filter with weights or weight function h_m. Substituting Equation 4.28 into Equation 5.15 yields

$$\Gamma_X(f) = \Delta t \sum_{k=-\infty}^{\infty} \left[\sum_{m=0}^{\infty} \sum_{n=0}^{\infty} h_m \, h_n \, \gamma_Z(k - n + m) \right] e^{-i2\pi fk\Delta t}$$

$$= \Delta t \sum_{m=0}^{\infty} h_m \, e^{i2\pi fm\Delta t} \sum_{n=0}^{\infty} h_n \, e^{-i2\pi fn\Delta t} \sum_{k=-\infty}^{\infty} \gamma_Z(k - n + m) \, e^{-i2\pi f(k-n+m)\Delta t} \tag{5.17}$$

which reduces to

$$\Gamma_X(f) = \Gamma_Z(f) \left| \sum_{m=0}^{\infty} h_m \, e^{-i2\pi fm\Delta t} \right|^2, \quad -1/(2\Delta t) \le f \le 1/(2\Delta t). \tag{5.18}$$

Equation 5.18 shows that the input and output spectra of a linear filter are related to each other through the modulus squared of the Fourier transform of the weight function h_m. Applying Equation 3.14, we get

$$\Gamma_X(f) = \Gamma_Z(f) |H(f)|^2, \qquad -1/(2\Delta t) \le f \le 1/(2\Delta t). \qquad (5.19)$$

$H(f)$ is the impulse or frequency response function originally defined in Section 2.6. The square of its modulus is appropriately called the *variance transfer function* in that it describes the transfer of variance from the input spectrum of a random process that passes through a linear filter to the output spectrum.

Now consider the general linear process discussed in Section 4.3 in which the input process is white noise. From Equation 4.30 $\gamma_Z (k - n + m) = 0$ for all values of the argument, except when $k = n - m$, in which case $\gamma_Z(0) = \sigma_Z^2$ and

$$\Gamma_X(f) = \Delta t \, \sigma_Z^2 \, |H(f)|^2, \qquad -1/(2\Delta t) \le f \le 1/(2\Delta t). \qquad (5.20)$$

$\Gamma_X(f)$ in Equation 5.20 is the variance density spectrum for a general linear process. The variance transfer function serves to shape the output spectrum given an input spectrum that is uniform with frequency. The output spectrum is a consequence of the weight function applied to the white noise input.

5.4 Spectra of selected processes

5.4.1 White noise

In Section 4.4.1 we found that for the case of white noise, $h_0 = 1$ and all other weights are zero. Therefore, $|H(f)|^2$ in Equation 5.20 has unit value and the variance density spectrum is

$$\Gamma_X(f) = \Delta t \, \sigma_Z^2, \qquad -1/(2\Delta t) \le f \le 1/(2\Delta t). \qquad (5.21)$$

Integration of Equation 5.21 over the entire frequency range of the spectrum yields the population variance, σ_Z^2.

5.4.2 First-order autoregression

The formula for a first-order autoregression or AR(1) was given in Section 4.4.2 as

$$X_t - \mu = \alpha(X_{t-1} - \mu) + Z_t \qquad (5.22)$$

but by expanding backward in time it can be written also in the form of the nonrecursive filter (Equation 3.5)

$$X_t - \mu = \sum_{m=0}^{\infty} h_m Z_{t-m} \qquad (5.23)$$

where $h_m = \alpha^m$, $m \geq 0$.

Now let us introduce a complex sinusoid

$$Z_t = e^{i2\pi f t \Delta t} \qquad (5.24)$$

in place of Z_t in Equation 5.22. Observe that subscript t is a time increment counter and, as such, is dimensionless. Firstly, consider Equation 5.22 to be, temporarily, the working formula for calculating a realization, in which case the notation would be lower case. Then, think of z_t to be any sinusoidal component of the white noise input. How its output amplitude changes as a consequence of the recursive filter is a function only of frequency.

From Equation 5.23, the output is

$$X_t - \mu = \sum_{m=0}^{\infty} h_m e^{i2\pi f(t-m)\Delta t}. \qquad (5.25)$$

Expanding Equation 5.25 yields

$$X_t - \mu = e^{i2\pi f t \Delta t} \sum_{m=0}^{\infty} h_m e^{-i2\pi f m \Delta t}$$

or

$$X_t - \mu = e^{i2\pi f t \Delta t} H(f) \qquad (5.26)$$

where $H(f)$, as noted earlier, is the impulse or frequency response function from Equation 3.14.

The result of substituting Equation 5.26 into Equation 5.22 is

$$H(f) \left[e^{i2\pi f t \Delta t} - \alpha\, e^{i2\pi f(t-1)\Delta t} \right] = e^{i2\pi f t \Delta t} \qquad (5.27)$$

or

$$H(f) = \frac{1}{1 - \alpha\, e^{-i2\pi f \Delta t}}. \qquad (5.28)$$

In line with previous statements, we can think of the white noise process Z_t as being comprised of sinusoids, so that Equation 5.28 provides the response at any frequency f. Therefore, from Equation 5.20, the variance density spectrum for an AR(1) process is

$$\Gamma_X(f) = \frac{\Delta t\, \sigma_Z^2}{\left|1 - \alpha\, e^{-i2\pi f \Delta t}\right|^2} \tag{5.29}$$

or

$$\Gamma_X(f) = \frac{\Delta t\, \sigma_Z^2}{\left[1 + \alpha^2 - 2\alpha \cos(2\pi f \Delta t)\right]}, \qquad -1(2\Delta t) \le f \le 1/(2\Delta t). \tag{5.30}$$

An example of a variance density spectrum for an AR(1) process is given in Section 5.4.4.

5.4.3 Second-order autoregression

The equation for the second-order autoregression or AR(2) process was given in Section 4.4.3 and is

$$X_t - \mu = \alpha_1(X_{t-1} - \mu) + \alpha_2(X_{t-2} - \mu) + Z_t. \tag{5.31}$$

As shown by Jenkins and Watts (1968, p. 228), for an AR(2) process

$$H(f) = \frac{1}{1 - \alpha_1\, e^{-i2\pi f \Delta t} - \alpha_2\, e^{-i4\pi f \Delta t}} \tag{5.32}$$

with the result that from Equation 5.20 its variance density spectrum is

$$\Gamma(f) = \frac{\Delta t \sigma_Z^2}{1 + \alpha_1^2 + \alpha_2^2 - 2\alpha_1(1 - \alpha_2)\cos(2\pi f \Delta t) - 2\alpha_2 \cos(4\pi f \Delta t)}, \tag{5.33}$$

$$-1/(2\Delta t) \le f \le 1/(2\Delta t).$$

Obtaining Equation 5.33 is a rewarding exercise and the last of seven we have recommended to enhance your understanding of various equations and other developments that have been presented throughout this book.

In problem 5 at the end of this chapter you are asked to compute the population variance density spectrum of an AR(2) process and the sample variance density spectrum of the realization of an AR(2) process you computed in problem 9 of Chapter 4.

5.4.4 An example of first-order autoregression

For the example in this section, we return to Figure 4.9, which shows a realization of an AR(1) process with $\rho(1) = 0.9$. Our goal is to compute and interpret the variance density spectrum of this highly serially correlated time series. The first step toward this goal is to calculate the sample acvf using Equation 4.11, the result of which is the heavy solid line in Figure 5.4. We observe a strong peak around 40 lag units. If we mentally fit a smooth curve to the time series in Figure 4.9, we see peaks at point 0 and in the neighborhood of points 40 and 80, and troughs in between. This oscillation is accounted for in the acvf by the peak around $k = 40$ (peak-to-peak distance in the time series) and the trough around $k = 20$ (trough-to-peak or peak-to-trough distance). Similarly, the trough in the acvf around 55 corresponds to the peak in the time series around point 0 and the trough around point 55 and the peak around point 40 and trough around point 95. There seems to be more than one oscillation in the times series. You may recall from problem 9 in Chapter 4 that the acvf of a time series that is a sinusoid is itself a sinusoid with the same period as in the time series.

We can compare the observed acvf discussed above with the population acf scaled to the variance of the realization. The dashed line in Figure 5.4 is the acf for the AR(1) process, Equation 4.42, multiplied by the variance of the realization or

$$\rho^k(1) \times 2.4652$$

where k is lag number. While we would not have expected the shape of the observed acvf to match the shape of the scaled population acf, it is surprising how unrelated

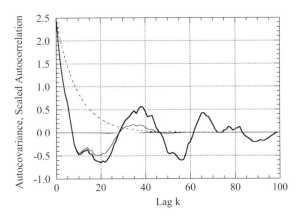

Figure 5.4   Heavy solid line: observed acvf of the realization of 100 values from an AR(1) process shown in Figure 4.9 where $\rho(1) = 0.9$. Dashed line: population acf $\rho(k)$ multiplied by acvf(0) = 2.4652. Light solid line: observed acvf multiplied by the Tukey lag window with maximum lag equal to 60.

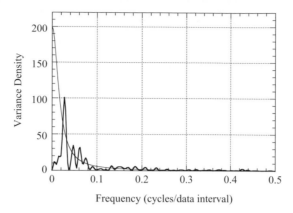

Figure 5.5   The heavy line shows the variance density spectrum (variance/(cycles/data interval)) of the raw or unwindowed acvf in Figure 5.4. The light line shows the population variance density spectrum from Equation 5.30.

they are. From experience, it appears that the greater the population autocorrelation $\rho(1)$, the greater the potential for a sample acvf to substantially deviate from the population acvf.

The heavy line in Figure 5.5 is the variance density spectrum of the time series in Figure 4.9 computed using Equation 5.9 multiplied by two to provide a folded spectrum. Stated differently, the spectrum is one sided and contains all the variance in the times series. The strong peak in Figure 5.5 occurs at 0.0275 cycles/data interval, which is equivalent to a data length of 36 units, which is in accord with both the time series in Figure 4.9 and the acvf in Figure 5.4. In addition, the spectrum exhibits other nearby peaks but lesser in magnitude, in agreement with the acvf.

We can compare also the sample variance density spectrum to its population spectrum given by Equation 5.30 and shown in Figure 5.5 by the light line. As with the sample and population acvfs, there is considerable disparity between the sample and population spectra. The area under the curve of the former spectrum is 2.4652, while that under the latter spectrum is 5.2632. The variance in the realization is less than one-half the population variance.

5.5 Smoothing the spectrum

In Section 5.1 we discussed the lagged product approach to smoothing "noisy-looking" periodograms and in Figure 5.1 showed two ways to achieve this. The procedure described in Figure 5.1a is the easier of the two. In this method, we multiply the computed acvf (referred to as unwindowed or raw) by a lag window to obtain a windowed acvf. The Fourier transform of the windowed acvf produces a smoothed spectrum. The smoothing procedure is discussed in this section.

As a first step we list desirable properties of the lag window, denoted by w(k). They are:

(1) $w(0) = 1$. Because the total variance in a realization occurs at zero lag in the acvf, this property preserves the total variance.

(2) $w(-k) = w(k)$. This property results in a spectrum window that is real only.

(3) w(lkl) decreases smoothly as lkl increases. This property helps to keep side lobes of the spectrum window small. As we saw in Section 3.4.1, the spectral decomposition of a rectangular function produced the Gibbs oscillation.

(4) w(lkl) becomes zero at some lag lkl < N. Apart from the shape of the lag window the maximum lag also controls the degree of smoothing of the raw spectrum.

From Equations 5.8 and 5.11, we conclude that the lag window and spectrum window form a Fourier transform pair such that

$$W(f) = \Delta t \sum_{k=-M}^{M} w(k) \, e^{-i2\pi fk\Delta t}, \qquad -1/(2\Delta t) \le f \le 1/(2\Delta t) \qquad (5.34)$$

where $M < N$ and

$$w(k) = \int_{-1/(2\Delta t)}^{1/(2\Delta t)} W(f) \, e^{i2\pi fk\Delta t} \, df, \qquad |k| \le M. \qquad (5.35)$$

To satisfy property (1) of the lag window, that is, $w(0) = 1$, we see from Equation 5.35 it is required that

$$\int_{-1/(2\Delta t)}^{1/(2\Delta t)} W(f) \, df = 1. \qquad (5.36)$$

From Figure 5.1a, the Fourier transform pair for smoothed spectra is

$$\overline{C}(f) = \Delta t \sum_{k=-M}^{M} w(k) \, c(k) \, e^{-i2\pi fk\Delta t}, \qquad -1/(2\Delta t) \le f \le 1/(2\Delta t) \qquad (5.37)$$

and

$$w(k) \, c(k) = \int_{-1/(2\Delta t)}^{1/(2\Delta t)} \overline{C}(f) \, e^{i2\pi fk\Delta t} \, df, \qquad |k| \le M \qquad (5.38)$$

which can be written also as

$$\overline{C}(f) = \Delta t \left[c(0) + 2 \sum_{k=1}^{M} w(k)\, c(k) \cos(2\pi f k \Delta t) \right], \qquad -1/(2\Delta t) \le f \le 1/(2\Delta t)$$

$$(5.39)$$

and

$$w(k)\, c(k) = \frac{1}{2M\Delta t} \sum_{n=-(M-1)}^{M} \overline{C}(f_n)\, e^{i2\pi f_n k \Delta t}, \qquad |k| \le M \qquad (5.40)$$

where $f_n = n/(2M\Delta t)$. Equation 5.40 follows from Appendix 5.B in which N is replaced by M.

An example of a lag window that possesses the desirable properties given earlier is

$$w(k) = \begin{cases} \dfrac{1}{2}[\cos(\pi k/M) + 1], & |k| \le M \\[2mm] 0, & |k| > M \end{cases} \qquad (5.41)$$

and is called the cosine window or Tukey window (after John W. Tukey, 1915–2000, a famous mathematician). It is the same as the von Hann filter that was discussed in Section 3.2.3, and its Fourier transform (Jenkins and Watts, 1968, p. 252) is

$$W(f) = M\Delta t \left[\frac{\sin(2\pi f M \Delta t)}{2\pi f M \Delta t} \times \frac{1}{1 - (2fM\Delta t)^2} \right], \qquad -1/(2\Delta t) \le f \le 1/(2\Delta t).$$

$$(5.42)$$

To understand the effect of spectrum smoothing, we can apply the Tukey window to the acvf in Figure 5.4 with, say, a maximum lag M of 60. The result is the light solid line in Figure 5.4. There is not much noticeable smoothing of the complete acvf until lag 10, after which the acvf is increasingly damped up to lag 60, where it becomes zero. The results of substituting the product of $c(k)$ and $w(k)$ from Equation 5.41 into Equation 5.39 and carrying out the computations are shown in Figure 5.6 by the heavy line. As in Figure 5.5 the variance density spectrum in Equation 5.39 has been folded. For comparison, the light line shows the raw or unwindowed spectrum from Figure 5.5. Also, to provide greater separation between curves, the resolution of the vertical axis has been doubled relative to that in Figure 5.5. We observe immediately the smoothing effect of multiplying the acvf

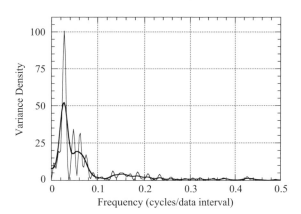

Figure 5.6   The heavy line shows the variance density spectrum of the windowed acvf in Figure 5.4. For reference, the light line shows the unwindowed variance density spectrum in Figure 5.5. Note that the vertical scale here has twice the resolution of that in Figure 5.5.

by the Tukey window before transformation (or the convolution of Equation 5.42 with the light line). The principal peak has been almost halved and the adjacent peaks and troughs almost completely smoothed. The area under each curve (or variance) is identical.

We can easily surmise that had we selected a smaller maximum lag M, say 30, there would have been even more smoothing. In practice, the amount of smoothing one should apply is somewhat arbitrary. If we think of a spectrum comprised of signal plus noise, the goal would be to design a spectrum window that smooths the spectrum in such a way that the noise is reduced but the primary features of the spectrum are preserved. Another approach might be that we wish to smooth the sample spectrum so that it better characterizes the population spectrum. Unfortunately, we practically never know the underlying spectrum. Even if we thought we did, we would conclude from the example just studied that there can be great divergence between the population and sample variance density spectra. A lengthy discussion of spectrum smoothing can be found in Jenkins and Watts (1968, pp. 274–284).

Appendix 5.A   Proof of Equation 5.11

The goal of Appendix 5.A is to prove that

$$c(k) = \int_{-1/(2\Delta t)}^{1/(2\Delta t)} C(f)\, e^{i2\pi fk\Delta t}\, df, \quad |k| \leq N - 1. \tag{5.11}$$

We begin with Equation 5.8:

$$C(f) = \Delta t \sum_{k=-(N-1)}^{(N-1)} c(k)\, e^{-i2\pi\, fk\Delta t}, \qquad -1/(2\Delta t) \le f \le 1/(2\Delta t). \tag{5.8}$$

Now substitute Equation 5.8 into Equation 5.11 and interchange the order of summation and integration. The result is

$$c(k) = \int_{-1/(2\Delta t)}^{1/(2\Delta t)} \left[\Delta t \sum_{n=-(N-1)}^{(N-1)} c(n)\, e^{-i2\pi fn\Delta t} \right] e^{i2\pi fk\Delta t}\, df$$

$$= \sum_{n=-(N-1)}^{N-1} \Delta t\, c(n) \int_{-1/(2\Delta t)}^{1/(2\Delta t)} e^{i2\pi f(k-n)\Delta t}\, df$$

$$= \sum_{n=-(N-1)}^{N-1} \Delta t\, c(n) \int_{-1/(2\Delta t)}^{1/(2\Delta t)} \cos[2\pi f(k-n)\Delta t]\, df$$

$$= \sum_{n=-(N-1)}^{N-1} \Delta t\, c(n)\, \frac{\sin[2\pi f(k-n)\Delta t]}{2\pi(k-n)\Delta t} \Bigg|_{-1/(2\Delta t)}^{1/(2\Delta t)}$$

$$= \sum_{n=-(N-1)}^{N-1} c(n)\, \frac{\sin[\pi(k-n)]}{\pi(k-n)}$$

$$= \sum_{n=-(N-1)}^{N-1} c(n)\, \delta_{kn}$$

$$= c(k) \qquad \text{QED.}$$

The function δ_{kn} is the Kronecker delta in which

$$\delta_{kn} = \begin{cases} 1, & k = n \\ 0, & k \ne n \end{cases}.$$

Appendix 5.B Proof of Equation 5.12

The goal of Appendix 5.B is to prove that

$$c(k) = \frac{1}{2N\Delta t} \sum_{n=-(N-1)}^{N} C(f_n)\, e^{i2\pi f_n k\Delta t} \tag{5.12}$$

where the summation is over the frequencies $f_n = n/(2N\Delta t)$. Notice that the frequency spacing is one-half the usual spacing associated with a periodogram of length N. At these frequencies we have, from Equation 5.8,

$$C(f_n) = \Delta t \sum_{k=-(N-1)}^{N-1} c(k)\, e^{-i2\pi f_n k \Delta t}. \tag{5.B.1}$$

Now substitute Equation 5.B.1 into Equation 5.12 and interchange the order of summation. The result is

$$c(k) = \frac{1}{2N\Delta t} \sum_{n=-(N-1)}^{N} \left[\Delta t \sum_{p=-(N-1)}^{N-1} c(p)\, e^{-i2\pi f_n p \Delta t} \right] e^{i2\pi f_n k \Delta t}$$

$$= \frac{\Delta t}{2N\Delta t} \sum_{p=-(N-1)}^{N-1} c(p) \left[\sum_{n=-(N-1)}^{N} e^{i2\pi(k-p)n/(2N)} \right]$$

$$= \frac{1}{2N} \sum_{p=-(N-1)}^{N-1} c(p) \left[\sum_{n=-(N-1)}^{N} e^{i\pi(k-p)n/N} \right]$$

$$= \frac{1}{2N} \sum_{p=-(N-1)}^{N-1} c(p) \begin{cases} e^{i\pi(k-p)/(2N)}\, \dfrac{\sin[\pi(k-p)]}{\sin[\pi(k-p)/(2N)]}, & k \neq p \\[2ex] 2N, & k = p \end{cases}$$

$$= c(k), \quad \text{since } \sin[\pi(k-p)] \text{ is zero for the case of } k \neq p. \quad \text{QED.}$$

The penultimate equation is obtained using Equation 1.B.4 for the sum of complex exponentials.

Problems

1 List two reasons why one might wish to compute a variance density spectrum of a time series via the autocovariance function as opposed to computing a periodogram.

2 List the four desirable properties of a lag window and explain why each is important.

3 We derived the equation

$$\Gamma_X(f) = |H(f)|^2\, \Gamma_Z(f)$$

where H(f) is called the frequency response function.

(a) What is $|H(f)|^2$ called?

(b) If the variance density spectrum $\Gamma_Z(f)$ is white noise, write down the equation for $\Gamma_X(f)$ in terms of the variance of the white noise process.

(c) The equation for $H(f)$ for an AR(1) process is $H(f) = 1/[1 - \alpha \times \exp(-i2\pi f\Delta t)]$. Derive the equation for $\Gamma_X(f)$ for an AR(1) process reduced to its most utilitarian form.

4 Write a computer program that will compute a smoothed variance density spectrum using the formula

$$\overline{C}(f) = 2\Delta t \left[c(0) + 2 \sum_{k=1}^{M-1} w(k)\, c(k)\, \cos(2\pi fk\Delta t) \right], \quad 0 \leq f \leq 1/(2\Delta t)$$

where the notation is as usual.

   In order to test the correctness of your spectrum, include in your program the computation of the total variance from all the positive frequencies in the spectrum and show that the total variance in the spectrum is equal to the variance computed directly from the data. In summing the spectrum variance, make sure that you use a one-half bandwidth at $f=0$ and $f=1/(2\Delta t)$.

5 (a) Using the program you developed in problem 4, compute the sample variance density spectrum for each of the four time series described in Problem 9 of Chapter 4, using the same maximum lag and the Tukey lag window. Let $\Delta t = 1$ s and compute spectral estimates at increments in frequency of 0.005 Hz.

(b) Plot the one-sided (total variance) sample spectrum and population mean spectrum for each of the four time series on linear axes. For the analytic sinusoidal time series the "population" mean spectrum would be a delta function at the appropriate frequency with the appropriate area. Put the sample and population mean spectra for each time series on one graph so that they can be easily compared and include the respective total variances on each graph.

(c) For the AR(1) and AR(2) cases, verify that the population mean variance computed directly from the process (square the process formula and take its expectation) is the same as the variance in the population mean spectrum that you plotted in (b).

(d) Discuss each sample spectrum in relation to the associated time series. For example, is its form or shape in agreement with the structure of the time series?

(e) Discuss each sample spectrum in relation to the population mean spectrum. For example, is it essentially coincident with the population mean spectrum or are there substantial departures?

References

Jenkins, G.M., and Watts, D.G. (1968) *Spectral Analysis and its Applications.* Holden-Day, San Francisco, CA.

Koopmans, L.H. (1974) *The Spectral Analysis of Time Series.* Academic Press, New York.

Tennekes, H., and Lumley, J.L. (1972) *A First Course in Turbulence.* MIT Press, Cambridge, MA.

Index

Time Series Analysis in Meteorology and Climatology: An Introduction, First Edition. Claude Duchon and Robert Hale.
© 2012 John Wiley & Sons, Ltd. Published 2012 by John Wiley & Sons, Ltd.